WITHDRAWN
UTSA LIBRARIES

THE AGRICULTURAL BLOC

BY
ARTHUR CAPPER

GREENWOOD PRESS, PUBLISHERS
WESTPORT, CONNECTICUT

The Library of Congress has catalogued this publication as follows:

Library of Congress Cataloging in Publication Data

Capper, Arthur, 1865-1951.
  The agricultural bloc.

  Reprint of the 1922 ed., issued in series: The Farmer's bookshelf.
  1. Agriculture--Economic aspects--United States.
  2. Agriculture--United States. I. Title. II. Series.
  HD1761.C36   1972   338.1'0973   78-136848
  ISBN 0-8371-5282-8

Originally published in 1922
by Harcourt, Brace and Company, New York

First Greenwood Reprinting 1972

Library of Congress Catalogue Card Number 78-136848

ISBN 0-8371-5282-8

Printed in the United States of America

EDITOR'S PREFACE

Not since the so-called "Granger movement" of the late '70's has there been so definite and widespread an agrarian movement in the United States as at present. The American farmers during the period from about 1895 to 1915 enjoyed in the mass a considerable degree of prosperity, although during those years there was a gradually growing feeling of unrest, rooted in the belief that the farmer was increasingly the victim of economic injustice. When we entered the war, the farmers soon came to feel that they had no voice in arranging matters that affected them, and that the men in charge of the larger affairs had inadequate knowledge of the farmers' problems or sympathy with their point of view. The post-war deflation affected the farmers more seriously than probably any other class of our people, in fact, so seriously as to be all but disastrous.

Meantime the organizations of farmers had been increasing in power. The Grange had grown steadily. The Farmers' Union in some parts of the country had become quite powerful. The coöperative movement had made greater progress in the last half dozen years than perhaps in any two decades of our previous history. The rise of the Non-partisan League indicated that the farmers were prepared to resist, and even to fight politically, in order to obtain what they regarded as reasonable justice. Out of this general situation grew almost over night what is at present the largest and most powerful of our farmers' organizations, the Farm Bureau Federation.

Recently the pressure from all of these organizations has been for legislation to relieve the situation, and the "Agricultural Bloc" was the result. This book describes the causes, methods, and results of this political aspect of the present agrarian movement. The author has been identified with the group from the start. He comes from one of the great agricultural states, is a successful editor of farm papers, and is in close touch and sympathy with the agricultural leaders as well as with the rank and file.

The "Agricultural Bloc" may or may not

be a passing phase. Comparatively few farmers care for or believe in a farmers' party; but the farmers are determined to seek their rights. Whether they become aggressive continuously, whether they seek to maintain permanent political groupings, depends largely upon how they are treated. As Senator Capper says, a farm group should not be necessary. The American people should have such an understanding of the farmer's position and problems and such sympathy with his point of view, as to make an agrarian movement unnecessary. One cannot avoid the reflection, however, that historically the rural people have been either neglected or exploited, or the further remark that, the world over, there is at present more rural unrest than in any recent period of history. It may also be observed that in Europe particularly, the peasants have more political power than they have ever had before.

Consequently this book, while it treats of a particular phase of the agrarian movement in the United States, is after all dealing with the symptoms of something fundamental and widespread.

KENYON L. BUTTERFIELD.

CONTENTS

CHAPTER		PAGE
I.	What is the Agricultural Bloc?	3
II.	The Crisis in Agriculture	13
III.	The Farmers' Part in the World War	25
IV.	The After-war Depression	38
V.	The High Cost of Living and the Farmer	48
VI.	Deficiencies in Farm Finance	57
VII.	The Burden of Transportation	69
VIII.	The Problems in Marketing	79
IX.	The Struggles of Coöperation	92
X.	Protection for Agriculture	105
XI.	The Public Attitude toward the Bloc	118
XII.	The Farmers' Program	131
XIII.	The Program of the Bloc	140
XIV.	The Record of the Bloc in Congress	153
XV.	What the Future Demands	162

THE AGRICULTURAL BLOC

CHAPTER I

WHAT IS THE AGRICULTURAL BLOC?

THE Agricultural Bloc is that group of Members of Congress who have recognized that an economic upheavel threatens the Nation if the foundations of our prosperity are weakened.

They recognize that American welfare depends upon the land and upon a permanent and prosperous agriculture.

They recognize that the American farmer not only feeds and clothes us but that he is also the best customer of our industries.

They recognize that the American farmer supports commerce, foreign and domestic, by producing over half of our exports; pays over half of the annual cost of transportation and maintains more than half of our public institutions.

They recognize that national prosperity is dependent primarily upon agricultural prosperity and that unless that third of our population who live upon the farms prospers the

Nation cannot have a continued growth and development.

They recognize that we have passed into a new era in our national history in which we cannot allow the balance of real production which comes only from the land to get out of balance with the dependent manufacturing industries, commerce, banking and government.

They recognize as just the claim of the man on the land for an equal voice in national affairs with those who live by trade, banking and manufactures which deal only with advancing stages of the raw product provided by the land.

They recognize the American farm as the nursery of a genuine freeborn citizenship which is the strength of the Republic.

They recognize as a menace to the national welfare any economic, social or political change which threatens to weaken that part of our citizens who live and work upon the land.

They recognize that an agricultural interest cannot in the end be a selfish class interest since nature allows no monopoly in agriculture. Better food production benefits all; commerce is increased; trade is stimulated; banking must expand and progress is promoted as the people on the land increase in affluence and net revenue

WHAT IS THE AGRICULTURAL BLOC? 5

which must be spent for the products of the labor of city workers.

The organization of the agricultural group in the Congress should not have been necessary. Since the foundation of the United States the fixed national policy has been to foster the opportunity of the man on the land. Our earliest pioneers came to America because of the opportunities to live in freedom on the land; land that has become more and more the foster-mother of liberty. Even while enjoying the greatest growth ever experienced by any nation, a large part of our people have drifted away from the primary principle that the interests of agriculture must not be submerged under the interests of industrial and commercial development. While professing great faith in the man on the land our people have developed an apathy toward the real needs of agriculture until an impending disaster, long forecast and foretold, had to break upon our heads to awaken us.

Even then, the apathy continued. For three whole years an almost constant series of warnings by our leading statesmen, economists and thinkers failed to awaken the nation to the need.

Then the American farmer asserted himself.

He became convinced that he must stop the drift toward the shoals of indifference. He recognized the power of organization and there grew, almost overnight, large and determined groups, under the leadership of seasoned men who were farm-bred and trained. Finding mere appeal brought scant results, these leaders took council with their chosen representatives in the houses of Congress.

Let us review some of the irritating factors in the agricultural situation which has stimulated Congress to give closer attention to agricultural needs. Crops had been sold at a loss, when marketable at all, because of the high cost of production and excessive freight rates. The factors involved in the cost of production of farm crops, principally labor and interest on capital involved, were still at high levels though the price of the finished product had dropped far below the level of the pre-war period.

Freight rates, of which the farmer pays considerably over half of the total, had not been reduced to any extent in 1921, and were having a paralyzing effect upon production. Farmers were forced to decide whether they should change their system of farming, as high freight rates would ultimately force them to do, or whether they might expect freight rate reduction within a reasonable time.

The dollar value of crops produced on American farms had practically doubled during the war, consequently the requirements for money and credit to finance the production and movement of these crops had also doubled, and with cost of production still high the credit needs of agriculture were far above the limited facilities afforded in our antiquated credit system which had been developed primarily to foster commerce and trade rather than agriculture. The increased cost of doing business influenced the distributing and marketing agencies to compete with farmers for credit, and either left the farmer to carry the burden of the surplus or dump it upon an already declining and congested market.

The hazards of farming as a business, which must always be a serious matter, were increased because of the lack of business organization among farmers and as a matter of self preservation the farmers of the United States turned to self organization to dispose of their products or to control them while they were held in elevators and warehouses until consumptive demand require that they be put upon the market.

This organization movement met opposition from business concerns, both large and small, and the very organization which was designed to promote orderly marketing to stabilize prices and conserve the food supply for the consumer was declared in conflict with the anti-trust legislation.

Efforts to secure a definition and authority for farm coöperation under the law were met with opposition from middlemen's organizations and large business concerns. Farm coöperation, which has proved successful for generations in Europe and other countries, was strenuously opposed here. There was not an economist nor an authority on agriculture who did not advocate coöperation to some degree. They all knew that the methods of a corporation would not do for agriculture because farms are independent units, each widely separated unit producing a variety of products, that can not be welded into a corporate aggregation.

For many years the cattle men had protested against the control of the live stock industry by the packing industry, and there had been constant agitation for packer control but without result. The abuses of the grain exchanges had been pointed out repeatedly without any-

WHAT IS THE AGRICULTURAL BLOC? 9

thing being done about it. Other measures long advocated had received no serious consideration.

The farmers of America found themselves being opposed instead of aided, by business groups which should be the best friends of agriculture. All of these influences worked together to create the unanimous feeling among agricultural leaders that the nation must be brought to realize the danger of further delay on these questions.

Early in the present administration it became evident to several Senators and Representatives from agricultural districts that a definite plan of procedure must be agreed upon in order that the economic needs of agriculture might be carefully studied and remedies proposed which would be effective.

The agricultural group, later named the "Bloc" by popular writers, was formed at a meeting called by Senator Kenyon, May 9, 1921, held at the Washington office of the American Farm Bureau Federation, at which twelve Senators met to decide on a program for immediate attention. It included an equal number of representatives of the two leading political parties, principally Senators from the

Middle West and South, our great agricultural sections, where the situation was most acute. Those present were Senators W. S. Kenyon, of Iowa; Arthur Capper, of Kansas; G. W. Norris, of Nebraska; F. R. Gooding, of Idaho; E. F. Ladd, of North Dakota; Robert M. La Follette, of Wisconsin; E. D. Smith, of South Carolina; J. B. Kendrick, of Wyoming; Duncan U. Fletcher, of Florida; Joseph E. Ransdell, of Louisiana; J. T. Heflin, of Alabama, and Morris Sheppard, of Texas.

At this meeting also there were present representatives of government departments, asked in to act as advisers on the program that should be adopted. There were representatives of the farmers who were asked to tell what farmers felt was their outstanding needs. The declaration of purpose by Senator W. S. Kenyon, of Iowa, was that this group give thorough and earnest consideration to the outstanding proposals to the end of securing action by Congress. Four Committees were appointed on the following subjects: Transportation, Federal Reserve Act, Commodity Financing and miscellaneous agricultural bills.

From the very beginning this movement was non-partisan and a recognition of the economic

WHAT IS THE AGRICULTURAL BLOC? 11

crisis; an endeavor to outline a plan for an economic re-adjustment rather than a scheme to gain partisan advantage. It declared *for* things rather than *against* them; for harmonizing views, not for creating discord; for coöperation, not antagonism; and for all citizens, not for farmers alone.

The outstanding reason that brought this group together was the fact that the general public and the majority of Congress had not realized that the nation had passed into a new economic era in which the balance between agriculture and other industries must be more carefully safeguarded. From this beginning in May, 1921, the group was enlarged to include some 22 Senators and meetings were held from time to time at the office of Senator Kenyon.

From the very first Senator Kenyon was recognized as the leader of the group and it was chiefly due to his sincere interest in agriculture and vigorous leadership that the Bloc became effective.

The following Senators joined the group: Charles L. McNary, of Oregon; Peter Norbeck, of South Dakota; John W. Harreld, of Oklahoma; A. A. Jones, of New Mexico; William J. Harris, of Georgia; H. F. Ashurst, of Arizona;

Pat Harrison, of Mississippi; Wesley L. Jones, of Washington; Robert N. Stanfield, of Oregon; Frank B. Kellogg, of Minnesota; Charles A. Rawson, of Iowa, and Claude A. Swanson, of Virginia.

A similar movement was started in the House and a group of Representatives with the same purpose and non-partisan objective was organized to represent the leading agricultural districts. The group has not been so thoroughly established as in the Senate, however, nor has it functioned in such an effective manner. Agricultural matters have usually received more attention in the House than in the Senate.

The Agricultural Bloc was the result of the conviction forced upon the more thoughtful representatives of the farmers that they must unite on a simple and direct program in order to bring the nation to see the needs in the emergency and to act before it was too late.

CHAPTER II

THE CRISIS IN AGRICULTURE

GREAT movements have many forms and find expression in various ways. The beginnings are not clearly defined at the outset, and it often takes a long view and a perspective of years to observe the significance of a great movement, whether it be economic, social or political. Sometimes a mere incident serves to bring the big purpose into view or place it in the foreground and this incident becomes the center of public interest, though it is but a part of the greater chain of important events that are occurring.

The great economic upheaval of the world, resulting from the World War, has had a tremendous reaction upon industries which has been most strikingly shown in its effect upon agriculture, the basic industry. The great changes that have rocked the foundations of centuries of standing and sent tremors through the whole economic structure also shook the foundations of American agriculture, long regarded as the impregnable base of America's prosperity.

The movement was not primarily political nor social, but economic with social and political reactions. Even fundamental economic principles of long standing have been freely questioned and the entire basis of our industrial organization subjected to renewed criticism. When any great movement of this character occurs it is reflected immediately in the thought of large groups of people who discover themselves in a new set of circumstances which they have not before experienced, which cause them to question all established institutions and frequently to produce new and radical departure from principles that have been tried and proven true through history. Under the stress of sacrifice and pain men are driven to radical extremes. The result is great discontent, intolerance for established law and order, disdain for the tried rules of orderly industry, and a general economic upheavel that affects all industries.

At such a time sane and cool-headed statesmen stand fast against the waves on the surface of the storm of criticism and endeavor to seek fundamental causes and to uncover real sources of discontent. Then it is futile to endeavor to remedy a disease until the diagnosis has been thorough. The hopelessness of satisfying popular clamor without getting at a basic need and the futility of legislation for emergency purposes only, is not generally understood without studying the fundamental weakness of existing laws and procedures to the end of securing permanent improvement.

When the reaction in trade following the war brought the most precipitate and crushing depression ever experienced by American agriculture, a widespread feeling of discontent resulted. There was immediately reflected in the halls of Government, as is every great change at some time or another, the fact that American agriculture demanded special attention.

The American farmer arose to the war emergency in a manner which astonished the world, even those who had followed his tremendous growth through recent years. Within a space of two or three years production was adjusted to the rapidly growing demand of Europe for war emergency purposes and the call upon the farmers to aid in winning the war by feeding Europe was answered in most positive terms.

Then, with even greater suddenness, occurred the reaction and the depression laying bare the real weaknesses in our industrial system, bringing us again to realize that agriculture is the basic industry in this country and that when it suffers, all industry suffers with it and in an increasing degree as the agricultural depression continues.

When this crisis in American agriculture began to be reflected by appeals to the American Congress, through the Representatives of the regions particularly distressed, the true situation was presented by them in the two Houses of Congress with great emphasis. But the country's distress was not confined to agriculture alone. The public and most members of Congress were not yet ready to seek for the real causes and much of the protest and appeal from agricultural representatives went unheeded.

The farmer's complaint was dismissed as merely that of a group in temporary need, with the unsatisfying counsel that they wait the normal turn of events. When it was pointed out that the farmers had been patient during the war, withholding complaints for known injustices for patriotic motives, they were met by the rebuff that others too had suffered and any remedy must be universal in application rather than to be applied to any specific industry. The more the farmers insisted, the more there developed a criticism that a selfish class interest was involved rather than one that was fundamental to prosperity in all lines.

From the time of the Armistice through 1919 and 1920 there was a growing and more determined protest from farmers and agricultural leaders that things were not right with the farm industry and remedial measures on a national scale must be attempted. Our older observers compared this protest with that which occurred succeeding previous wars and were inclined to dismiss the matter as merely a repetition of the unrest that follows in the wake of war.

But American agriculture was in a new position. The period of expansion in new areas was practically at an end. Further expansion must be made by improvement in efficiency on the part of the individual worker on the land and not merely through expansion in numbers or in areas.

For ten years statisticians had been noting the change in the trend of affairs in that Ameri-

18   THE AGRICULTURAL BLOC

can agriculture was showing the first signs of not keeping pace in its growth with the national requirements. The temporary spurt in production experienced during the war, while showing what might be expected for a brief period, was made at such a cost that it could not be expected to continue as a normal growth later on. It was a growth accomplished by sacrifice of the best methods for anything that would bring quick results.

The United States has been a country not only self-sufficing, but an exporting country with regard to food-stuffs. While we were dependent upon some neighbors for the food products not grown here, they were much more than offset by the volume of our export, except in very recent years.

An authority presents this matter in a few concise figures. In 1860 there were 13 acres of farm land per capita of population, while in 1910 this area had dropped to 9.5 acres per capita, thus in 50 years there was a reduction which left only 73 per cent as much farm land for each person in the United States as there had been 50 years before. Not only had there been a decrease in the number of acres, but the average quality of the land in farms had deteriorated because of the inclusion of lower grade land in the western plains and mountain counties.

From another point of view, the change is emphasized even more strongly. In 1880 70.5 per cent of the population of the country was rural, while in 1920 53.7 per cent was rural. The number actually on farms was not even this large, since the latter census included villages and towns of less than 2,500 inhabitants of which there were a large number. The farm population in 1920 was roughly about one-third of the entire population of the country.

For fifty years the momentum of development was in agriculture; the subduing of new and fertile land by a constant increase of immigrated settlers had been keeping production on an increase. But more recently this stream of new workers had been passing into industries and the growth of agriculture had been occurring through the improved efficiency of the farmer, due to the increased use of machinery and more scientific practices in farming. But both scientists and economists know the limitations of these factors in improvement, and the student of world history was beginning to issue warnings of the impending decline of agriculture, such as has occurred in nearly every great civilization which has disappeared.

American agriculture had been developed on a basis entirely different from that of any preceding great nation because from the outset the American farmer has been a citizen in equal standing and opportunity with the merchants, the traders and the manufacturers. This meant that a continued growth of commercial and industrial America, if accomplished by reaching out to distant countries for food, would be accompanied by the decline of the American farmer to a lower grade of living. In most other countries the farmers have been held in a peasant class, subservient to a ruling commercial class.

The American farmer has been peculiarly jealous of his large part in the development of the Republic and from the outset we have regarded equal opportunity and chance of achievement as a cornerstone of our growth. Consequently, when conditions arise that forecast a decline in agriculture the American farmer is prompted to immediately interpret it as a reaction which will affect the nation as a whole.

The rapid commercial and industrial growth

THE CRISIS IN AGRICULTURE 21

of the United States during the past 25 years has been reflected by the growing national interest in these forms of development. A constructive long-time program of development for agriculture had been discussed at times, but comparatively little has been done about it. Following the Civil War there was great impetus given to the development of the application of science to farming through the establishment of the Federal Department of Agriculture, the State Agricultural Experiment Stations and later the State Colleges.

Scientific farming was hailed as the means by which all of the needs of agriculture would be met and during a period of 50 years the advances made in this field have been tremendous. From a beginning when there was practically no organized science of farming we developed rapidly until we were recognized as the most progressive of all nations in our methods.

Likewise, improvements in agricultural machinery came with astonishing frequency and resulted in greater production from an increased efficiency of the relatively small number engaged in productive agriculture. The business of farming was organized and systematized until the American farm had become a business unit with as high efficiency as had been obtained in manufacture and commerce.

From time to time, however, foresighted statesmen had observed needs which had not been provided for and special surveys were made to uncover the true situation. When the Country Life Commission was appointed, due to an appreciation by Theodore Roosevelt of the fundamental importance of agriculture and its relation to the national welfare, there were many skeptics who felt that the surveys were unnecessary and even Congress refused for a time to recognize the merit of the conclusions which were developed by this Commission. This served to give a nation-wide viewpoint to many, however, who had heretofore been living in relatively limited circles.

Still later, the rapid growth of agriculture as an industry, demanding a larger volume of invested and working capital, forced attention to be given to the needs of rural credit and a commission was appointed to investigate systems followed in older countries with a view to developing a financial system which would meet the needs of American farmers. This commission brought about the development of the Federal Land Bank system which was just beginning its operations when the World War began in Europe.

While these surveys had given an impetus to the movement for a national policy for agricultural development, the results were piecemeal and by no means fully rounded out into a progressive program. The decade from 1910 to 1920 witnessed the development of many national advances looking to a more stable agriculture, both in the Federal Government and in the States. But the program of progress was largely set aside when the United States entered the war.

Representation for agriculture in Congress has been a matter of concern for several years and the war served to intensify the feeling among farmers that though their representatives were disposed to give a fair consideration to new proposals, when it came to a matter of conflict, agricultural interests were likely to be those last considered. During the war, there occurred a succession of incidents which intensified this feeling. Following the war, when the first indications of a coming crisis were to be noticed, agricultural leaders hastened time and again to Washington with warnings and appeals that something be done to establish a proper balance between the agricultural industries and other industries.

Through 1919 and 1920 Congress was being continuously implored to give special attention to the needs of agriculture with but scant hearings being granted. Naturally the proposals were many and varied and lacking in organized direction. This situation continued until the election of 1920 passed, but increased with renewed vigor following that election, thereby showing that the agitation was not a partisan, political matter, but based upon a real need.

CHAPTER III

THE FARMERS' PART IN THE WORLD WAR

AMERICAN farms provided the substance necessary to the winning of the war in the threefold nature of men, food and money. Enough time has passed since the world war to give us a fair perspective view of the supreme test which agriculture underwent in this period of stress.

The war began for American farmers with the outbreak in 1914 even though the full strain of the emergency did not culminate until we were joined in the conflict in 1917; that is, great economic influences began to bear which had much to do with the situation which has followed. The stimulation of manufactures, due to the demand for munitions to European nations then in conflict, immediately placed a draft upon labor which was felt on the farm. The stimulant of advancing prices for foodstuffs led the American farmer to increase his production and strive to take advantage of the opportunity afforded by an expanding market. He, therefore, began to work under high pressure, not realizing that when the United States should enter the war later he would be expected to continue this increased rate of endeavor as a means of aiding our own country in the conflict.

The growth in American farming as indicated by the total acreage in 15 principal crops is shown in the following table:

1910	290,000,000 acres
1914	300,000,000 "
1915	309,000,000 "
1916	310,000,000 "
1917	321,000,000 "
1918	350,000,000 "

While the increase was steady but slow during the period preceding 1917, amounting to about 3½ per cent from 1910 to 1914 and 6 per cent from 1914 to 1917, the great increase came in 1918 after our entry into the war.

The increase in production of some of our principal crops from the average of 1910 to 1918 was not as great as the increase in acreage would indicate. The corn crop of 1918 was a little less than that of the pre-war period. The wheat crop while amounting to 917,000,000 bushels was less than the great crop of 1915 and only about 200,000,000 bushels above the pre-war crop.

While farmers could expand their operations to more crop acreage by plowing up their pastures and using land which normally would have been left alone, they could not control the yield factor, which, with respect to some crops, was none too favorable during the war years. There was a shortage of fertilizers in some quarters which also tended to reduce production. The farmers did their best at the outset by planting an increased acreage whenever possible.

The efforts to increase live stock production in the time of war could not be expected to have great results when the demand was so general for breadstuffs. Live stock producers were apprehensive concerning prices and so long a time was required to mature live stock that they hesitated to go into it in an emergency manner. It takes more time to build up a live stock business than it does to launch into crop production. But in spite of the various handicaps there was an increase in live stock kept in the United States though not equal to the increased demand of Europe for live stock products.

The number of horses on farms increased

slightly, as did the number of mules. Milch cows increased in numbers considerably, other cattle the greatest of all, and swine to an even greater extent than the larger animals. Sheep alone declined in numbers.

The increase in exports of animal products from pre-war average to 1918 was a striking feature of the situation. Beef of all forms was exported during the period of 1910-1914 to the extent of 220,000,000 pounds, while in 1918 exports were over 502,000,000. Pork and lard exports during the pre-war period amounted to about 900,000,000 pounds, while in 1918 the total was more than 1,900,000,000. Our exports of condensed milk alone increased from a pre-war average of around 15,000,000 pounds to 530,000,-000 pounds in 1918.

The war began to affect agriculture first through the withdrawal of surplus farm labor which was attracted to the war industries by the rapidly increasing wages that were offered. When we entered the war it was already clear that the increased program of production required of the farmers would call for an unusual amount of farm labor. There immediately followed the organization by the Government of campaigns to secure for food production, men above military age, others unable to enter the army on account of physical defects, and boys not old enough for army service.

Those in charge of the food stimulation campaign did not seem to realize, however, that it is not easy to draft large numbers of untrained people into farm work and get the best of results, and while great aid was rendered by the services of outside and inexperienced workers the results from this farm labor campaign were by no means so great as some had hoped that they would be.

It was estimated by the Department of Agriculture that at the beginning of the war about fourteen million men were engaged in farming. Before a year had passed about one and a half per cent of these had been drafted and in addition to this a large number had joined the army as volunteers and probably a much greater number had gone into the industries because of higher wages. While the draft regulations were drawn with a view to taking from the farms relatively a smaller number of men, this had little effect since public sentiment did not excuse the able-bodied farm boy who remained at home, however necessary his services were on the farm. Then, as always, the vigorous lads from the farms were those who wanted to be in the front of the fight and in most cases they would not claim or accept exemption when it was offered to them.

While the efforts to stimulate the removal of labor from cities to the farms were widespread, the results were not extensive when counted in numbers compared to the total number engaged in farming. In fact, the Department of Agriculture estimated that probably a total of around 100,000 boys who had not before worked on farms were employed at some time during the season of 1917 while a somewhat larger number helped with the harvest in 1918.

The fact was that farmers continued and increased their field of production without asking for outside help, largely by doing more work themselves with the aid of their families. There can be no measure of the sacrifices made by fathers and mothers to replace the hands of sons who went into the army.

Farm wages began to increase considerably in 1917 and a year later were about double what they were before the war. The fact that there was no serious calamity due to labor shortage on farms was due more to the increased efforts of those who were left on the farms than to any addition of help from the cities. There was a good deal of popular talk of how city men and boys went into the country and saved the harvest and how business organizations formed groups of men to work in the fields, but their real value in helping the farmer was comparatively small.

The work was speeded up more by the ingenuity of the practical farmer in using machinery and the farm women and children than by all help from the outside. All this speeding up increased the farmer's costs. Not only was it more expensive to keep more horses and tractors, but horse feed, gasoline and oil cost more and the burden laid upon the farmer was exceedingly heavy.

Added to the handicaps of shortage of labor and higher cost of machinery was the growing scarcity of fertilizer which was a serious factor in the eastern states and with special crops like potatoes. The excessively high prices of fertilizer added to the cost of production even more than the increased cost of labor in many instances.

In the midst of the widespread campaign for increased production which farmers were asked to make in the face of increasing costs came the fixing of the price of wheat. The handling of this matter by our government is not a pleasant story when viewed from the farmer's standpoint.

The fact that food control was imposed with so little protest was due to the genuine patriotism of all the people concerned and particularly the farmers. When the committee on prices was appointed by the President and reported that the price of wheat should be placed at $2.20 for Number 1 Northern Spring Wheat at Chicago, farmers knew at once that many would be unable to produce wheat at this price and get back the cost of production, but they continued without a murmur. The fact that some of the farm leaders pointed out that this could hardly be a fair price was made the occasion for unwarranted criticism that the farmers were trying to profiteer on the increased food production.

While the amount that the farmer actually received for his wheat in 1917 was considerably more than received during 1916, the increase in costs was so much more that the dissatisfaction continued to grow. Farmers believed and expected that the crop of 1917 would bring a higher price than they had received in 1916 before the outbreak of the war and to a certain extent repay those who had had short crops for two years previously. Their complaint was not that they should be guaranteed a profit at the expense of the bread consumers, but that the costs of production had been considered in connection with other industries by the adoption of the cost-plus system which enabled other industries to pay high wages and still maintain their usual margins of profit. If cost-plus was fair for manufacturers they contended that it should be fair for farmers.

When the subject of wheat control came up in 1918, Congress made an effort to raise the price of wheat to $2.40 per bushel by passing a bill to that effect but it was vetoed by the President. The increase in freight rates which had occurred in the meantime, since the crop of 1917 had been marketed, was presented as another argument why a better price should be offered farmers for wheat.

This agitation over the price of wheat would have probably passed without serious effect upon farmers had not the charges of profiteering and hoarding developed in the city press. The answer to these hoarding charges is now clearly shown by the statistics of wheat held by farmers on March 1 of several years, which was as follows:

	Million Bushels
1910	160
1912	122
1914	152
1917	101
1918	111

It will be seen that the amount held in the war years, 1917 and 1918, was much lower than the average of previous years. In fact, farmers scraped the bins before the harvest of 1918, bringing the surplus down to only eight million bushels left in farmers' hands—a dangerously low surplus.

The efforts at stabilizing pork production were somewhat less irritating to farmers and the complaint arising from this situation was due more to the indefiniteness of the proposition than to the actual relation that was fixed between the price of hogs and the price of corn. The Government made fewer mistakes in the handling of the live stock situation than in some other lines. It did, however, discourage the production of highly finished beef at a time when the market was not demanding a large amount of light-weight beef. The result was a serious break in the prices of cattle and a very large and unnecessary loss to cattle growers who sold light cattle at times at scarcely more than their hides were worth. Then meatless days were provided for and consumers had to pay unusually high prices for beef. It is undoubtedly true that our live stock industry suffered more acutely from depression later because of the manner in which this beef control was handled. However, the farmer proceeded to increase the supply of meat at a time when a shortage was feared. We exported large quantities to the Allies and thereby aided in prosecuting war.

The control of dairy products, milk, butter and cheese, was attended by considerable disturbance which was exaggerated by the press denouncing dairymen and dairy farmers as profiteers and the almost constant agitation for legal proceedings against milk producers' associations. The same coöperative organizations which had been heretofore hailed as a great improvement in marketing development were subjected to a constant fear of prosecution and an agitation that prices were not fair to the consumer. Milk producers were indicted and while they were sincere in their belief that they were helping improve the situation, they were constantly harassed by public criticism.

The charge of profiteering was so generally

applied to producers as to create more and more unrest. We are now able to view the price relations from a disinterested standpoint and while there was a great increase in agricultural prosperity in terms of actual dollars received for the crop, this must be considered in relation to the necessary expense of production. The index of the price of farm products in 1918 was 111 per cent above that of 1915 while the cost of the bulk of things purchased by farmers had increased about 200 per cent. The wholesale prices of all commodities had been constantly above the price of farm products from 1915 to 1917. For a short time in late 1917 and the early months of 1918, farm prices ran higher than wholesale prices. Then again, wholesale prices advanced and have since been constantly above farm prices.

This charge of profiteering by farmers, coming at a time when they had strained every nerve to keep production on the increase, was a most bitter experience because of its unfairness and particularly because the popular press seemed to lack all appreciation of what a factor food production had been in winning the war. The increase in land prices which came later was seized upon as another opportunity to show why the farmer has gained an undue profit. Yet this increase was received by comparatively few and was expensive to all because it advanced the basic values used for taxation appraisals.

CHAPTER IV

THE AFTER-WAR DEPRESSION

DURING the period of high prices in 1919, when corn was worth one dollar and a half a bushel, a farmer could get five gallons of gasoline for a single bushel. A year later at the farm price that bushel of corn would buy only one gallon of gasoline; and two years later the same bushel of corn would buy about half a gallon of gasoline.

In 1919 it took only six bushels of corn to buy a ton of coal; while a year later it cost 40 bushels of corn and two years later, 60 bushels of corn.

Likewise in the period of high prices, 40 bushels of corn, or the product of a fair acre, would buy a $60 suit of clothes. A year later it took 200 bushels of corn to buy the same suit, and two years later 300 bushels.

Such comparisons as these show how real was the depression in the purchasing power of the American farmer during the period following the era of high prices. Considering the price

THE AFTER-WAR DEPRESSION 39

levels alone, the crisis was not so apparent at first, because many farm prices were still above the level of 1913, and in fact they have so continued in many instances until the present.

The difficulty arose from the fact that farm prices declined earlier and more suddenly than other prices, thereby bringing a hardship on the farmer because of his reduced purchasing power.

The farmers' dollar was worth less than in 1913 for several months during 1916 and 1917, which was a period commonly supposed to be a very prosperous time for farmers. The reason that it was not prosperous for farmers was because general wholesale prices increased more rapidly than farm prices.

The real depression began in January, 1920, at a time when the country as a whole was enjoying almost unprecedented prosperity and was in a disposition to give scant attention to warnings or complaints from any source. The farmers' dollar in January, 1920, was worth a dollar in purchasing power in other products than food and farm products. Then followed a steady decline until December when this same dollar was worth only 65 cents and the average for that year was 86 cents. It was during 1920 that Congress heard its first warnings and the needs of agriculture were first presented by those who were awake to the situation.

The decline in the farmers' dollar continued through 1921 without abatement and the bottom was not reached until December of that year, when the first signs of improvement could be noticed. It was in May, 1921, that the agricultural group first began to study the situation carefully, since even then there had been a steady decline of a year and a half and it did not require great foresight to see what was coming. This decline in agricultural prices was the most violent drop ever experienced, because prices had risen to heights never before equaled in our history, and within two years had dropped to low levels seldom before experienced. This was not a new economic event since historical statistics show that there were similar declines following previous wars, followed by periods of slow recovery. Had this situation been recognized by the government in sufficient time, there is no question that the blow could have been softened, thousands of bankruptcies avoided and untold suffering stayed.

But we learn only by experience and it seems that each generation must undergo its own har-

THE AFTER-WAR DEPRESSION 41

rowing experiences. The few warning voices raised early in the after-war period were given scant attention.

In the reports of the Joint Commission of Agricultural Inquiry, on the Agricultural Crisis and its Causes, there is to be found abundant testimony of what occurred. This report fortunately provides a permanent record of the changes in agriculture during the war, and may help to avoid another similar experience in the future.

The Commission endeavored to present the condition of the farmer: First, by a discussion of the purchasing power of his products. Second, a comparison of absolute prices of farm products with actual prices of other groups of commodities. Third, the quantity production of agriculture and other industries. And, Fourth, the reward for capital and labor employed in agriculture compared with other industries.

The report of this Commission is available to all who are interested in this subject, but the main conclusions are reviewed here because the facts concerning the agricultural prices must be kept in mind in order to understand this new movement among farmers. The decline in the purchasing power of the farmers' dollar has been referred to and can be best epitomized in a brief table as follows:

PURCHASING POWER OF FARM PRODUCTS IN TERMS OF OTHER PRODUCTS (U.S.D.A. INDEX)

Years	
1913	100
1915	106
1917	106
1918	112
1919	111
1920	86
1921	67
1922 to June	71

The price changes in other commodities lagged behind prices of farm products in the decline and retail prices lagged even further behind wholesale prices. Farm prices came down first, and other prices slower and in no instance have they yet descended to the low levels reached by farm products. It was this widening margin between the prices of farm products at the farm, and the prices which the farmer paid for what he had to buy, that caused the hardship. Retail margins during the time when farm products were at their lowest level were still far above the margins of 1913 as they still are in 1922.

The costs of distribution and various services have stayed on a new and higher level and there can be no full relief to agriculture until farm

THE AFTER-WAR DEPRESSION 43

prices again rise considerably above the pre-war normal or other prices come down.

It has been stated and believed by many that the American farmer brought on his own destruction by overproduction following the war. There was a continuation of a campaign to boost food production long after the war was over under the pretext of helping to revive the starved people of Europe. Production was stimulated artificially to a considerable degree. Yet the total production of grain was only slightly larger in 1920 than the average of 1909 to 1913, while the production of cotton was relatively less. The number of cattle and sheep on farms in 1920 was less than pre-war five-year period, although the number of hogs was somewhat greater. The slaughter of cattle and sheep was below the pre-war average.

The public was largely led to wrong conclusions by the greatly increased dollar value of crops produced, estimated on the basis of prices of 1919 and early 1920, rather than the production figures. This emphasizes a fact which has been the cause of much misunderstanding of the farm situation, namely the regarding of the dollar value of a crop as the true measure of its size and as a measure of the return to the farmer. The total value of a crop can never be estimated by multiplying its quantity by its market price at any given time, because the crops of the year are harvested at one period, but are consumed over 12 months. The live stock are counted but once a year while the consumption extends over the entire year and the question of average prices is involved, which never can be known until after the year has passed.

Overproduction was not the cause of the decline in farm prices, nor was the opportunity for foreign markets a serious consideration.

What of the farmers' income before and during the war? After a comprehensive survey of the facts the Commission of Agricultural Inquiry concluded that profits made by the farmer during the war were only slightly greater than those of 1913 and were swept away by the decline in prices of 1920 and 1921. It must be noted that when the Commission reached this conclusion the decline had not yet ceased and even greater losses were experienced by farmers during the latter months of 1921. While the Commission found that agriculture has produced normally about 18 per cent of the national income, the return received by the farmer for

THE AFTER-WAR DEPRESSION 45

his labor, after deducting 5 per cent on his property investment, was below the return received by employees in other industries. The average reward per farmer for labor, risk and management, after allowing 5 per cent on his investment, was as follows:

```
1909 ..................... $  311
1918 .....................  1,278
1920 .....................    465
```

Compare with this the reward of employees in the mining industry, which were in 1909, $590, in 1918 $1,280. Railway employees received in 1909 an average of $773 and in 1918 $1,532. It will be seen that the farmer would have been better off had he been working as a miner or for the railroads, so far as his return for labor was concerned, excepting his returns upon his savings represented in his investment.

The Commission concluded that while the income and reward for capital invested and labor employed in agriculture have been improving in recent years, as compared with other industries, they are still relatively lower than the rewards in other lines of business, and measured by the standard of the purchasing power of his products, the farmer was relatively worse off than those in other industries.

The question has been asked: Why is it that farmers continue in the industry so long as it is relatively less profitable than other industrial lines. The answer is, they don't.

The changes in the personnel on the farm, changes represented in movements of farm tenants, the steady decline in the percentage of workers engaged in agriculture compared to total population, are all indications that the business is not maintaining itself relative to the nation's business as a whole. This decline in workers in agriculture has not yet resulted in a serious shortage of foodstuff to the nation because of the possibility of expansion under pressure.

The Agricultural Commission found that measured in terms of increase of population, that is increase in production per inhabitant, food production has barely kept pace with consumption. Wheat and grains have increased in quantity about with the growth in population, but the total of live stock has been lagging. The decline in sheep is outstanding. The number of cattle other than milch cows was less than during the preceding decade. On the whole the consumption of meat is decreasing and we are gradually changing from a meat-eating to

THE AFTER-WAR DEPRESSION 47

a bread-eating nation. While the total number of milch cows has increased, the number of cows per thousand of population has decreased. While the production of milk sold has decreased there has been a revolution in the dairy industry so that the production of butter and cheese has declined, the milk being needed for consumption as whole milk rather than in the form of its manufactured products.

On the whole there has been nothing reassuring in the last decade in agricultural development from the national standpoint. It might be assumed that this shrinking of agriculture as shown by reduction in numbers of farms in relation to population, might mean a larger field for the American farmer to sell his product. But on the contrary, he is faced with a declining net return.

CHAPTER V

THE HIGH COST OF LIVING AND THE FARMER

THE subject of the high cost of living was uppermost in the minds of every one during the year 1919. The efforts to find a remedy for the high cost of living disease were directed mainly at the reduction of prices of foodstuffs. Most of these efforts resulted merely in the depression of the market for the products of the farmer and the live stock producer, while at the same time there was an increase in the cost of living for city consumers. Or in other words, the spread between the producer and the consumer was increased at the very time when it should have been decreased. The economic law appeared to work backwards. And while the market grew worse for the producer, prices went higher for consumers.

The public generally assumed that the farmer as a producer was not affected by the increases in prices of necessities. But this was a mistake. As soon as the control of the Food Ad-

HIGH COST OF LIVING AND FARMER 49

ministration was lifted, the prices of grain, by-products and mill feeds advanced immediately and this increased the cost of milk and pork production, at once laying a burden on the producer. Then came dollar-an-hour wages, which required about $3 wheat for the average wheat grower to break even.

This period of universal complaint about the high cost of living made it exceedingly difficult to work out adjustments between the industries. We could not continue to take away from producers the comparatively small profit and expect them to go on producing and increasing their output. Such a policy would lead inevitably to a decreased food supply and suffering among consumers. While it was logical for many city people to jump at once to the conclusion that the price of farm products be first reduced, this was a fundamental error.

It was my duty and opportunity early in my service in the Senate to examine into the consumers' side of this cost of living question. I was forced to the conclusion that the chief contributing element in our high living costs was our complicated and intricate system of distribution. While a wheat farmer had been forced to accept much less than the Government-guaranteed price for his wheat, and while live stock producers had incurred immense losses, and in many instances financial bankruptcy, the consumer has at the same time paid higher prices for nearly all food products. It was shown in 1919 that the only food items in which there were reductions in price to any extent, were navy beans, plate beef, chuck roast, and cornmeal. And the declines in these were less than 10 per cent, with the exception of beans, where the decrease was 30 per cent. In contrast to these declines, all other commodities in the list of staple foods showed increases, ranging from 1 per cent for round steak, to 28 per cent for lard and 85 per cent for onions. While the bottom fell out of the hog market, declining 8 cents a pound in 60 days, bacon went up 11 per cent above the war prices of the previous year.

The fact is, the increasing margin between producer and consumer was cutting both ways like a two-edged sword, causing suffering on the part of both groups. This is the way that it usually works, and illustrates a fact that the nation needs to realize, namely, that any condition that becomes oppressive to producers is sooner or later reflected in an increased cost to consumers, if not in actual shortage of food.

HIGH COST OF LIVING AND FARMER 51

Some specific figures will serve to show how real were the injustices that reacted on farmers during this period. Within a few weeks, the drop in the live stock markets in the central west cost cattle and hog raisers between $80,000,000 and $100,000,000 in decreased values. The Kansas cattlemen saw their animals drop in value $17 a head in two days. Hogs dropped $3 per head in value in a single day and three days later $10 to $12. When we reflect that it has taken six months to two years to fit these hogs and cattle for market, it is clear that such declines mean bankruptcy to the farmer who does not own his land clear and has a limited financial reserve.

The inequalities in prices of things was an injustice which was anything but encouraging to farmers. He knows that even if the price of his raw product is cut in two the consumer hardly notices the difference because of the long and increasing line of fixed profits and cost between producer and consumer. It takes four and a half bushels of wheat to make a barrel of flour. When the wheat raiser was getting $8.37 for the wheat, the miller got $12.50 for the flour, the baker $58.70 for the bread, and the hotel keeper selling the bread in slices in Washington received $587. It was stated in a hearing that even if the flour was given to the baker his bread would still cost 7 cents a loaf.

While the price of a single pair of shoes may keep one person in bread for a year, the hide of a steer a year and a half to two years old supplying enough leather for six or eight pairs of $12 shoes, brought the farmer only about $5. A pair of calfskin shoes frequently cost more than the farmer got for the live calf, hide and all. Somebody in between got what is paid for the veal, while the calfskin alone would make several pairs of shoes. These inequalities in prices and values in costs are due in many cases to increased middlemen's services which must in some manner be reduced if the real remedy is to be developed.

The farmer's attempt at the solution of these distributive problems, through the organization of better marketing systems under his own control, met with opposition from commission men and others who have sought to drive farmers from the field of distribution. When Ohio farmers who were selling milk actually below the cost of production attempted to practice collective bargaining, they were arrested and thrown into jail on the charge that they were combining in restraint of trade, yet what these farmers were attempting to do was to facilitate trade to increase the consumption of milk and to get a fair share of a consumers' price in order to foster production.

Our city people have overlooked the fact that it is not actual high prices from consumers that the farmer wants, but merely a price that bears a fair relation to the cost of production so that he may secure a reasonable margin. Farmers understand that a moderate price of the foodstuff will tend to encourage and increase production, thereby making it possible for him to sell more in quantity and get a larger gross return. He is not so much interested in what the actual price is, as in its relation to his costs and the prices of things which he has to buy. He fears the violent fluctuations of prices more than he does the long periods of relatively low prices. Farm production can be adjusted to any level of prices if the changes are not too frequent, and one of our greatest needs is for methods of stabilizing prices.

There has been considerable narrow thinking concerning the relation of the exportation of our surplus of farm products to the cost of living of our people at home. By providing an outlet for the surplus in foreign markets the American farm producer is encouraged to use all energy to increase production without being obliged to try to adjust the scale of production solely to American demand. Should farmers be forced to the latter course, the situation becomes hazardous for the consumer. If a surplus is to operate to break the price when a big crop is harvested, then the farmer will strive to produce just enough and no more. While the acreage put out to crops can be controlled, the yield to be harvested from that acreage may vary from 10 to 20 per cent, due to ordinary differences in seasons and weather. Hence, if the acreage is just enough to produce a nation's need with a given yield, and an unfavorable season cuts down the yield, there is a scarcity. The only safeguard for the public is in a surplus-producing agriculture and this is only encouraged when the farmers feel that the world's markets are open to them.

Much of the comment in the city press concerning the profits of farmers was based upon the reports of increases in land prices. The fallacy was widespread that such prices were beneficial to farmers in general, instead of an indication of a drift towards tenant farming, which was undesirable. The fact is, the extent of the increase in land values was much overstated, many of the reports being based upon the operation of land speculators who thrive in such a situation. Even farmers were sometimes misled and those who bought farms at record prices have since found that they are unable to pay for them. They have lost their investment and the land has gone back to the first owner since it has been impossible to make payments on the new high values with products selling at much lower values.

The city critics overlook the fact that any substantial and general advance in farm land prices is not desirable, even from the consumers' standpoint, since such advances merely increase the primary capital investment upon which an interest in turn must be paid out of the price received for food products and this increases the cost of production by a fixed charge that can not be ignored.

The cost of living agitation wore out in time and we have heard less of it since prices have begun to decline in all lines. But the evils of the wide margin between the prices paid to producers and those paid by consumers, continues; and unless we improve our methods of distribution, it is only a matter of a short time until another wave of discontent will arise over the cost of living. If wages in the industries were immediately reduced to-day to a level corresponding to the prices of farm products as they existed before the war, there would be a storm of protest even greater than that which we heard in 1919.

Popular complaint concerning the cost of living is based upon improper relations between income and the cost of necessities rather than the mere fact that prices are higher than they have been in the past. We gradually get used to new scales of prices and it is only a sudden change that brings out a protest.

The permanent solution for our cost of living problem centers in just three things: a more direct and economical marketing system; an efficient distributing system and vigilant oversight of trusts and credit. I am inclined to think that the greatest relief to agriculture and to the consumer will be obtained through cooperative marketing.

CHAPTER VI

DEFICIENCIES IN FARM FINANCE

THE high cost of money for agriculture has been the cause of increasing difficulty for farmers for a number of years. Denial of credit brought bankruptcy to many and serious troubles throughout the live stock regions. Deflation as it was brought about following the war reacted most disastrously upon the food producer, who is in the least advantageous position to help himself.

The deficiencies in our farm financial system were first recognized some fifteen years ago when the agitation for a better credit system, including land banks, personal associations and credit unions, began. As a result of several years discussion the Commission on Rural Credits and Coöperation was sent to Europe and after considerable study came back and made an extensive report. From this beginning there resulted the legislation which took the form of the Federal Farm Loan Act providing for the twelve Federal Land Banks.

The Federal Farm Loan System had just begun operations when the world war began in Europe. The preliminary organization was completed and loans began to be made available to farmers in increasing amounts, public confidence in the farm loan bank bonds was growing when opposition from land mortgage companies suddenly took form.

A test case was arranged and carried to the Supreme Court of the United States, the effect of which was to suddenly check the development of the system at the very time when its benefits were most needed because of the expansion that was taking place due to the better market for farm products. The case before the Supreme Court was delayed and for a year and a half the Federal Land Bank System stood still awaiting a verdict. Although the decision was favorable to the Banks and reassuring in the end, the delay was very serious because it came at a critical period when loans should have been made regularly.

When the Federal Land Bank System was inaugurated the need for some form of short-time rural credits was emphasized, but this feature was not added to the Land Bank System, principally because the Federal Reserve Act,

DEFICIENCIES IN FARM FINANCE

then a new proposal, had not yet been tested. It was believed that the Federal Reserve System would furnish to a large extent the flexible financial aid which is needed by agriculture for crop movements at harvest time. Though it was known from the outset that the Federal Reserve System was dominated by banking interests, directed by bankers, and was several steps removed from the country bank with whom the farmer deals, still it was hoped and expected by many that it would remove the need for a special system of rural credits for short-time personal loans.

The experience immediately preceding the war and during the few months following showed without question that prompt action must be taken by the Government. The call for aid from farmers to the banks was rejected and the country banks joined the farmers in appeals to Congress. Country banks were obliged to close their doors because they could not collect from farm debtors who were unable to pay because of the slump in prices in farm products which within a few short weeks had dropped to one-half what they had been. In addition to the inability of the banks to collect loans outstanding they were faced by a new crop year's calls from farmers for funds to meet current expenses and additional funds with which to plant a new crop.

This acute situation forced the marketing of breeding animals by the stockmen in an effort to liquidate their loans. Crops were dropped on the market at less than the cost of production, thereby forcing the decline in prices even faster. Early in the year the nation had been alarmed because of the prospect that farmers could not produce enough food. Credit was strained by the country bankers then in order to finance the farmers to put in a large crop. We were told that anarchy and bolshevism would result if the world was not fed and farmers were appealed to in every manner to aid the starving.

After this credit strain to produce a crop came the decline in prices and, of course, the result was inevitable. Country bankers were pinched and they in turn were obliged to press the farmer. Further up the line, the city banks were pressing their country correspondents for liquidation and on top the reserve system was pressing all for settlement. I pointed out to the Senate in December, 1920, that since June 1st of that year corn had slumped 70 per cent in

DEFICIENCIES IN FARM FINANCE 61

value and was down to about 30 cents a bushel at country shipping points. In November all records for shipments of cattle to the Chicago stockyards were broken when 4,503 carloads, containing over 111,900 head of cattle, were marketed and shipped in six days, and these shipments included thousands of breeding animals and calves which should have stayed on the farms to keep up the future live stock supply.

There was a wide-spread demand from farmers that the credit situation be eased by the Federal Reserve Board adopting a more liberal policy in rediscounting agricultural paper. The demand was not for funds for unsafe loans, and it was not just the hand-to-mouth farmer who was caught in the pinch, but the big substantial ranchers who in many cases produce the bulk of our live stock shipments.

The country as a whole appears to have ignored the agricultural credit situation and financial leaders talked deflation, the Secretary of the Treasury suspended the War Finance Corporation and no provision was made for even temporary relief until the situation had become almost unbearable. The Governors of several states in conference were among the first to join in the call for relief from Congress.

Had the credit needs of agriculture not been thoroughly canvassed just a few years before there might have been more excuse for the Administration overlooking this situation, but the matter had been fully agitated and it had been repeatedly shown that the ordinary deposit bank could not of itself take care of farm needs because credit for production and often for marketing must exceed the ordinary thirty to ninety day loan with which other business can operate. The farmer's period of credit varies greatly in the time he needs the money and he is often unable to know how long he will need the money when he makes the first investment.

When the Joint Commission of Agricultural Inquiry began its investigations into the credit situation, the depression was at its worst, and the testimony of those who appeared before the Commission reflected the hardships and losses most intensely. The Commission found some significant facts which should be considered in their logical order. The following excerpts present the conclusions:

Farm indebtedness has doubled in the last ten years and the drop in prices has the effect of again doubling this indebtedness.

DEFICIENCIES IN FARM FINANCE 63

These difficulties were due to the credit restrictions and limitations of the past 18 months and in part to the fact that the banking machinery of the country is not adapted to the farmers' requirements.

The Commission pointed out that the Federal Reserve System included about 8,210 national banks and only 1,630 of the approximately 20,000 state banks. The national and state banks are the principal agencies furnishing short-time credit to farmers and these, together with the farm loan system and the private farm mortgage companies, furnish the bulk of long time credit. Such short time credits as were available to farmers were largely limited to periods of six months or less owing to the fact that paper of longer maturity than six months for agricultural purposes is not eligible for rediscount with the Federal Reserve Bank. The Commission recommended that through adapting existing banking agencies a credit of sufficient maturity to make payment possible out of the proceeds of the farm should be provided. This means a credit running from six months to three years, depending upon the character of the crop or live stock to be marketed. In the case of crops, six months is often sufficient, but in the case of live stock, three years or more is necessary.

The Commission stated as its opinion that a policy of restriction of loans and discounts by advances in the discount rates of the Federal Reserve Banks could and should have been adopted in the early part of 1919, and that had this policy been adopted much of the expansion and speculation of the period following the war could have been avoided.

When finally the Federal Reserve Board adopted the policy of restriction of credits in 1920 expansion in currency and prices had reached such a point that deflation was accompanied by great losses and hardships.

The Commission believed that a policy of lower discount rates in greater liberality in extending credits could have been adopted in the latter part of 1920 and the early months of 1921, which would have retarded the process of liquidation and spread the losses over a longer period where they could not be avoided.

The growing demand for temporary relief brought about through the efforts of the Farm Bloc the revival of the War Finance Corporation which has served a highly useful purpose in providing credit and direct loans, saving

DEFICIENCIES IN FARM FINANCE 65

thousands of country banks and producers from bankruptcy. It has provided the means whereby many could continue production this year who would otherwise have been forced to suspend operations, move to less favorable farms and begin all over again.

The War Finance Corporation has been a most beneficial organization in many ways because of the fact that its Managing Director, Mr. Eugene Meyer, has from the outset given his time and ceaseless energy constantly to the problem of discovering ways by which the corporation could be of value to agriculture.

When the work of the corporation was suspended, following the close of the war, Mr. Meyer felt that the time was just coming when the corporation could be of real service to farmers. There were many who were interested in farm finance who had not thought of the corporation as the means through which the situation might be improved. Mr. Meyer came to Washington and held several conferences with the farm leaders. He also attended farmers' meetings elsewhere explaining how the corporation, if revived, might function. It is due to quite an extent to his convincing arguments that the corporation was given the opportunity to render the excellent service that it has since given to farmers.

Through the prompt organization effected by Mr. Meyer, money was immediately made available to farmers in amounts that are substantial in themselves and were the means of releasing even greater sums into general circulation.

Another important bill in the Farm Credit Series which was supported by the Agricultural Bloc was that which increased the interest rate on Federal Farm Loan bonds from 5 per cent to 5½ per cent without increasing the rate to farmers on loans. This made it possible for farm loan bonds to sell promptly in competition with other securities. Supplemented by the Curtis Bill supported by the Bloc, which provided for increasing Federal Farm Loan Bank capital by $25,000,000 for use as a revolving fund, the Bond-Selling bill became even more effective.

The proposal by the Farm Bloc that agriculture should have representation on the Federal Reserve Board was approved by leaders familiar with farm finance in all its phases. One authority in discussing the reason why a special representative of agriculture should be upon the Board pointed out that the credit needs

DEFICIENCIES IN FARM FINANCE   67

of agriculture differed from those of industry and commerce, moving more in seasons and cycles. There are many more hazards in agriculture which cause peculiar credit variation from year to year. Direct loans to farmers represented about 18 per cent of total bank loans and finally and by no means of least importance the members of the Federal Reserve Board find themselves soon growing out of immediate touch with the agricultural situation if they are not vitally connected with it by lively business interests.

When this bill was finally passed by Congress, there was immediately widespread interest among farmers in the selection of a man to fill the place, the farmers themselves being the most insistent that he be a man with a broad acquaintance with agriculture and not merely a farmer in name only.

The findings of the Joint Commission of Agricultural Inquiry pointed clearly to the desirability of meeting the need for additional rural credit by adapting existing banking agencies to the requirements, provided it is clearly set forth that the new form of credit will be provided by the new machinery.

This matter of short-time rural credit had

been discussed by so many authorities, however, that a number of plans were proposed by various Senators and Representatives in the form of bills.

The Agricultural Bloc held an open hearing on the question of short-time rural credit and heard representatives of the farmers and others discuss the situation. It was very apparent that all were interested in the result rather than the means, so a committee was appointed to consider the various bills proposed and arrive, if possible, at an understanding of the main points to be covered so that a new bill or a combination of bills might be introduced which the entire membership of the Bloc could support.

CHAPTER VII

THE BURDEN OF TRANSPORTATION

AGRICULTURE has always borne the larger part of the essential costs of transportation. This is due to the fact that the farmer cannot select his location entirely with respect to market, but must locate where the soil and climate will favor the production of the crop he chooses to grow. Only a few farmers can locate near centers of consumption, or ports of export. But there must always be a large part of our farm production that must be carried long distances from point of production to the centers of large groups of consumers.

The fact is oft stated, yet easily passed by, that the farmer pays freight both coming and going. His price for the goods that he sells is the wholesale price less the freight. While the cost of all that he buys is the actual cost plus the freight on the finished product to his farm. Included in the costs of the manufactured products which the farmer buys in large proportion are the freights upon raw materials which have been produced upon other farms.

Reliable and cheap transportation is essential to agriculture because it is only through the easy flow of products that consumption is encouraged and the market expanded. The cost of transportation to the farmer is not represented by freight rates alone, but by the effect upon his industry of the development of transportation systems of various forms, railroads, waterways and an ocean shipping system, which costs may be reflected in taxes or in a difference in the price of seed for the product he has to sell.

When prices increased with the beginning of the war, freight rates advanced slowly, and likewise since the period of high prices, they have remained high and are declining altogether too slowly. Experts agree that we should have a system of adjusting rates which will put them up when prices are high and the industries can stand a high cost of transportation and also make corresponding prompt reductions when prices decline. The tendency, however, when rates advance is for all interested parties, the railroads, the railway workers and the subsidiaries to the railroads, to keep rates at the higher level as long as possible.

The evil of the situation during the past few

THE BURDEN OF TRANSPORTATION 71

years has been due more to inequalities than to the mere fact that rates were high. Had they been high at the time other prices were high there would have been little complaint.

When railroads get more for hauling farm products than producers are paid for producing them, it is evident that rates are relatively out of line. Much of the depression and stagnation of business in 1921 was due to the fact that agriculture and industry was strangled by the grip of high rates on transportation. Sheep from our ranges, fruits from Florida and California, vegetables from southern gardens and even staple products could not be sold in the market for enough to pay the freight, consequently the producer was obliged to allow them to rot in the field rather than to try to ship these perishable products. Temporarily, of course, farmers kept on making shipments and producing with the hope that rates would soon be adjusted. But any permanent change in the cost of transportation means a decided change in the type of farming to be followed in regions a distance from market.

To show the effect of these rates on production of some of our products we have only to consider a few instances. First, there was the arrival at our seaports of shiploads of foreign products such as this country produces in exportable surplus. This became clear when we learned that a bushel of grain could be shipped from South America to New York for 12 cents while it cost 38 cents to ship a bushel of wheat from Minneapolis to New York. Another indication of the change in world trade is shown by the fact that cottonseed cake needed for feeding cattle in the middle western states could be shipped cheaper from Texas to Holland than from Texas to Kansas.

A car of grain shipped from the Texas Panhandle to market at an expense of $525 brought the producer $475. Texas and Florida truck farmers shipped many a carload of vegetables and received nothing in return except a bill for the balance due on freight and commission charges, and yet the products were not sold at excessively low prices to producers in the city market. A car of lettuce from a southern truck grower cost $491 for freight and brought $339 when sold. It cost more to ship California products than it did to grow them and in many instances the profit was more than wiped out.

The live stock industry was affected the same as crop production and cattle rates between

THE BURDEN OF TRANSPORTATION 73

main shipping points were out of all proportion to the value of the animal. Feeding stock from the ranges of New Mexico and Texas which are ordinarily shipped to Kansas to be finished on grass could not be moved since the large investments necessary to pay the high freights were held up by a shortage of funds in the banks. Loans for cattle feeding were almost impossible to secure. The price of the hay to be fed to much of this stock was likewise affected. Good alfalfa hay worth $6 to $8 a ton in Colorado cost $15 to $18 per ton to ship to Illinois, making the hay cost out of all proportion for its value for feeding purposes. The result was we were left with live stock without feed in one part of the country and feed producers found themselves without animals to consume the feed in other parts of the country with a barrier of high freight rates between them.

These rates affected the manufacturing industry by increasing the basic costs of raw material to a point where the industries in many instances chose to curtail production rather than pay the increased cost which they knew would drive the total cost of their product so high as to make it difficult to sell to the ultimate consumer. Iron mines had to choose between closing or operating at a loss, and the increased cost of the freight charges of handling ores and metals was reflected directly to the farmer in the higher costs of farm machinery.

Freight charges on coal alone increased as much as 200 per cent from 1914 to 1921 and it is reported that the average cost of transporting a ton of coal from the mines to the consumer early in 1922 was $2.74 while the price of the coal itself at the mines was $2.14. Coal is so essential to industrial activity that an increase of a cost factor such as this is reflected throughout the industry.

I was informed by a manufacturer of farm machinery that the freight bill on the material that goes into a grain harvester amounted to $35, a hay press $95 and a threshing machine $200 before the finished article starts on its way from the factory to the farmer. When we consider that it takes about six tons of raw material to make one ton of finished steel and three tons of raw steel to make one ton of farm machinery we can see how this item alone affects the price of the latter.

We will probably never fully know how large a part high freight rates played in the agricul-

THE BURDEN OF TRANSPORTATION 75

tural depression because of the fact that the increases in rates occurred almost simultaneously with a decline in prices. The experience of past years has been sufficient to warn us of the serious effects when rates become higher than the traffic will bear in the disturbances caused in food production and in the shifting of world trade and markets, all of which are expensive both to the nation and to the producer. Transportation charges are still too high.

It was believed when the war ended that the return of the railroads to private management would through the restoration of competition serve to promote efficiency and promptly adjust rates.

But our complex transportation problem had been neglected for some time and the war put us in a situation which required more heroic remedy. I proposed a bill early last year calling for the repeal of the Transportation Act insofar as it directed the Interstate Commerce Commission to make rates which would assure a return of 6 per cent to the carriers, because it seemed to me that readjustment has been retarded by the fact that a fixed return was included in the Transportation Act. It would appear to have been logical for transportation agencies to take losses along for a time with farmers, industries and other business in general and that the fixed return offers no permanent solution to our railroad problem.

The fact is that our railroads have developed under the private system with so much exploitation that the nation is now obliged to go to the bottom of the railroad problem, get at the facts and work out a new constructive program. This program must involve methods for adjusting rates in accordance with general business conditions and price levels so that they are relatively in line with values. There must also be developed a sympathetic attitude toward existing organizations and railroad leaders who in many instances are not responsible for the exploitation in the past but have inherited a problem of extreme difficulty. We cannot get along without our railroads because they are too much a part of our daily life for us to permit them to be crippled, and however much we regret the errors of the past, the future calls for development and expansion and increased efficiency.

It has been shown by the investigations of the Joint Commission of Agricultural Inquiry that the growth of a large number of systems and short line railroads under the com-

THE BURDEN OF TRANSPORTATION 77

petitive system has reached its maximum and that efficiency can only be obtained through consolidations of management and operating facilities. Railroads are essentially a public utility and all form a part of a national system. Any permanent improvement must come through making each part of the big unit perform its share of the work effectively.

All industries are alike interested in this problem. In the period of crop harvesting there must be the shifting of cars to the centers of grain production to handle that crop. When the need for fuel increases there must be opportunity to expand our transportation facilities for coal. All of this calls for an interchange of service between industries to take the place of the haphazard competition for cars at a time when a particular line of business is flourishing.

Transportation will be one of our biggest domestic problems for years to come. The old question of the highest efficiency of agriculture depends on adjusting rates to the character of the product hauled rather than to any flat tonnage basis. The development of export market, the use of a merchant marine and the encouragement of farmers to produce an exportable surplus, all hinge upon the question of proper adjustment of rates covering the transportation of products from our inland centers of high production to the shipping terminal.

The effect of these freight rates upon agriculture was not alone in the losses involved, but they also served to enlighten our agricultural people to their own interest in properly regulated transportation. Heretofore the farmer was seldom present when national policies regarding railroads were drafted. About the only time he appeared was to complain occasionally when rates got out of line with the value of some particular product. The individual roads would make an adjustment which would temporarily meet the demand and the disturbance passed over without any permanent remedy being developed. This system of adjustment merely postponed a day when the entire subject must necessarily be considered. We are approaching that day in surprising rapidity and there is nothing to be gained in postponing action further.

CHAPTER VIII

THE PROBLEMS IN MARKETING

AMERICAN agriculture has become over-burdened with an elaborate and wasteful system of marketing. The market manipulator and gambler has thrived in the market until the cost has become too great to bear. Both producer and consumer are suffering and the producer at present is suffering a little more than usual because of the readjustments. He parts with his products long before the consumer needs them and a host of toll and profit takers carry them in the meantime transferring them from hand to hand, absorbing as much profit as they are able to extract from the marketing business.

The farmer usually gets for his stuff about 30 per cent of the price which the consumer pays. Usually the least that the consumer is asked to pay is three to four times the farmers' selling price and the spread between the farmer and consumer will average about seventy cents of the consumer's dollar. In spite of the fact that this large margin cannot long continue to be taken, we have gone on increasing the number of middlemen.

The extremes to which these differences in farm price and consumer's price went last year can be easily illustrated. A mutton chop in a New York hotel cost more than was paid for a sheep in Colorado or Kansas. Ham is sold throughout the country at about six times the price of live hogs at Chicago when the normal ratio is supposed to be about one to three and one-half. They cannot understand how a campaign for greater meat consumption can succeed while this wide price range exists.

Take a case in reference to vegetables. A New Jersey farmer sold his potatoes to a commission house for $1.50 per barrel, receiving $9.00 for 6 barrels. After deducting the price of the barrels, freight and cartage and 10 per cent commission, the firm sent the farmer 99 cents for the potatoes.

It was shown during the investigation by the Senate Committee for the District of Columbia that a bushel of potatoes for which a farmer in Michigan received $1.00 was sold for from $4.00 to $6.00 in Washington and it passed through many hands. The farmer sold it to the village

THE PROBLEMS IN MARKETING 81

dealer who sold it to an agent of a Detroit buyer. The latter sold the potatoes to a commission agent in Washington who in turn sold them to the retail dealer, making in all six persons who had to receive a profit from handling these potatoes from producer to consumer. Well-developed coöperative marketing would probably eliminate at least four of these profit-takers.

I have elsewhere referred to the situation with regard to Texas cabbage which the Texas farmer sold for $6.00 per ton at his shipping-point and which brought $200 per ton at middle-western cities.

The public is not receiving the benefit of the sacrifices made by producers because of the large number of costs that are intervening between producer and consumer. High freight rates, of course, add greatly to the marketing and final cost to the consumer. These costs pile up when anything that is sold to other farmers is involved. When our dairymen in the south buy western hay for which the western growers get about $9.00 per ton, they pay about $24.00 a ton and then must endeavor to make milk at current prices by using such hay.

In addition to the many distributive costs

that are interposed between the raw product on the farm and the food on the consumer's table there is another large group of unnecessary influences. I refer to those who manipulate the market and speculate unnecessarily in food products, taking an unnecessary toll first from the producer and then from the consumer as the circumstances may change.

In July of 1920, the grain gamblers of the Chicago Board of Trade had very favorable circumstances in which to begin a great "bear" raid on the market which was maintained for nearly ten months. This was in the face of a large export demand for wheat. When the raid began wheat futures were selling at $2.75 per bushel but before it had ended the farm price of cash wheat had fallen to 85 cents. But during that time and for months afterward the consumer continued to pay war prices for bread and flour notwithstanding that the manipulation of the market had taken hundreds of millions of dollars out of the pockets of American grain producers. This experience was not only costly to the farmers but expensive to producers since the price of bread was not reduced as it might have been.

THE PROBLEMS IN MARKETING 83

Europe during the same year paid an average price of a dollar a bushel more for our wheat than the farmer who grew the wheat received in this country.

Representatives of the Board of Trade who appeared before the committees of the Senate and the House during the hearings upon the Future Trading Bill frankly admitted that market manipulation does go on. Abundance of proof can be found in the reports of the Federal Trade Commission showing to what amounts this trade in grain rises in some years. Three times as many bushels of grain as are produced in the whole world are sold in the Chicago market alone while the actual delivery of grain amounts to but a small per cent of the total transactions. Deals on the Chicago Board of Trade in 1920 amounted to 51 times the total amount of wheat produced in the United States.

It was a knowledge of conditions such as this that led me to introduce the bill which was known as the Capper-Tincher Anti-Grain-Gambling Bill which was passed by Congress in 1921, declared unconstitutional in some respects early this year and reënacted in new form by the House in June, 1922, awaiting attention in the Senate as this is written. Producers have been protesting against this form of speculation for years but it was not until the determination of the groups represented in the Agricultural Bloc forced the matter to the attention of Congress that results were obtained in the form of effective legislation. It is evident that American people have determined to do away with these methods of trading which have been expensive to every one concerned. Speculating in food supplies should not be necessary, as some claim it is, to carry over the surplus of one crop until the consumption needs of the people are ready to use it. If means of orderly marketing are needed, they should be supplied in a direct form through a better marketing system of combining insurance and warehousing which in the long run will be far less expensive to our people.

The fact is the route to market is too long and must be shortened and made more direct, and the charge for the services, where they are necessary services, should be distributed fairly.

The problem of the meat-packing industry, which has been constantly agitated for ten years, is a problem of fair distribution of charges and the elimination of manipulation of the retail

THE PROBLEMS IN MARKETING 85

market for meat. The packer question was discussed continuously before Congress for a long time and nothing was done about it until the Agricultural Bloc determined that there had been enough discussion and some action should be taken. Then within a few months, a bill was agreed upon, passed by Congress, tested in the courts, put into effective operation and is now declared to be an excellent piece of legislation. The consumer was really more interested in this packer legislation than the producer but he didn't know it. Consumers' organizations did not begin to help in framing a law until after the producers' representatives had determined that efforts should be made to get it passed without further delay.

As is true in many other cases, it has been the producer who has stimulated by his protests the public interest to the extent of definite action to work out a solution for many of these marketing problems from which the consumer will receive the greatest benefit in the end.

It is surprising to many that the one genuinely-prosperous nation in Europe to-day is little Denmark. The significant and encouraging fact is that this is the world's most thoroughly organized coöperative nation. Denmark is a country which is not naturally fertile, but through long years of successful coöperation the farmers of Denmark have developed not only their own industry but all industries in the country until they are truly prosperous. Their prosperity is sound because it is based upon good farming. It is the best indication of what coöperation will do for the consumer as well as for the producer since the benefits are reflected throughout the business of the nation.

One of the charges to the Joint Commission of Agricultural Inquiry was to study the marketing and distribution facilities of the country with a view to determining what the government might do toward improving conditions. This Commission has made one of the most thorough and comprehensive surveys of the situation that has ever been made and its report constitutes a classic among the literature on this subject.

The Commission has not arrived at the point of recommending any single remedy, legislative or economic, which will reduce the spread between producers' and consumers' prices because there is no single factor that is responsible for the trouble. Plenty of schemes have been proposed which were designed to solve the entire problem but they have never worked out. The

THE PROBLEMS IN MARKETING 87

solution depends upon the making of a number of smaller improvements from time to time as they can be adopted to advantage.

The conclusion of the Commission from its study of marketing and distribution should be carefully studied by every public man and by business men in all lines. It is only through a general understanding throughout the nation of these problems that we can expect to secure general support necessary to the adoption of reforms. The program of the Agricultural Bloc includes the enactment of measures that cover many of these points but chiefly those that are now before Congress and need immediate attention.

The Commission found that underlying all of these marketing difficuties is the need of better knowledge by producers of the processes through which their products go to reach the ultimate consumer. This must be supplemented by more information among retailers and distributors as to the requirements of consumers so as to encourage a continuous flow of raw materials into the markets for manufacturers and an equally steady flow of finished products through the channels of distribution to the consumer.

88   THE AGRICULTURAL BLOC

Some of the specific recommendations are as follows:

Standardization must begin with the producer so as to permit the selection and grading of products that will be of greatest value when they finally reach the consumer through the retail market. This part of the job of better marketing rests upon the farmer assisted by his investigators and educational agencies.

Uniformity of grades, standards, containers and measures which may be followed from the farm to the consumer's table will do much to eliminate waste and excessive cost. Legislation is necessary to establish the standards and to see that they are properly observed by every one concerned.

Coöperative organization among producers is the most effective means by which to put grades and standards into operation and to facilitate economical marketing. These organizations can adjust production to the requirements of the consumer and aid in avoiding the losses that come from over-production of foodstuffs of the wrong kind.

The arbitration of disagreements between shippers and receivers must be provided for through disinterested Boards supported by state

THE PROBLEMS IN MARKETING 89

or Federal legislation which provides them with power of action.

Warehouses to hold the production which is available at one time in the year, at harvest, and to take care of temporary surpluses to be distributed evenly throughout the year, are an agency which the consumer is as vitally interested in as the producer. Some of these warehouses should be at the centers of production while others can better be placed near the central wholesale markets.

Improved terminal markets in which the problems of grading and temporary storage can be handled before the products reach retail distribution are a necessity. Nearly all of our large cities have doubled in population since their terminal markets have been expended and the most congested points in the whole chain of distribution are to be found in these wholesale centers.

Orderly marketing in the broader sense is a necessity which we cannot continue to overlook. In some cases too extensive distribution has developed which involves excessive cost and disturbs the steady flow of farm products. In some instances more competition is needed while in others a smaller number of distributors might easily serve the needs with much greater economy. The outstanding example of the latter is the case of retail milk distribution. The duplication of milk deliveries when handled by several firms using almost exactly the same grades of milk are operating over the same territory.

One properly organized coöperative organization might easily serve milk to all of the people just as we have now in most cities a single water, gas, or electric light company, chartered to serve all the people under certain public regulations. When the product to be handled is exactly the same, there is nothing to be gained by multiplying the organizations engaged in distribution. When one agency, whether public or private, controls the distribution, however, it is necessary to have strict public supervision through legislation in order to protect the public interest.

It is difficult to see where our distributing system will end if it continues to grow as it has during the past few decades. Out of about 41,600,000 people, engaged in gainful occupation, 29,500,000 are engaged in manufacturing, transportation and distribution, or in rendering services other than actual production. Most of these services are necessary in our complex system of living, and we could not get along well without them, but there is no doubt whatever that some system looking to a reduction in the number of those who render no essential service must be devised. The best way to mend the present situation and provide for the future seems to me to be to increase the producer's profit by shortening the road to market. This will encourage production that is profitable, resulting in farm prosperity and greater prosperity in all lines of industry.

CHAPTER IX

THE STRUGGLES OF COÖPERATION

AMERICAN agriculture developed through the efforts of pioneers who were individualists, to a large degree. The free and open settlement of our western country developed large groups of men who were loosely associated for common defense but who were very independent when it came to matters of business. Probably no country has farmers as a class who have been less directed and controlled than those of the United States.

This individualism has made it difficult for agriculture to be organized as it has in some other older countries. There have been waves of interest in coöperation among farmers at various times in our history but these waves have later subsided and the period of coöperative organization apparently passed by for a time. In periods of stress farmers have been driven to realize the necessity of coöperation and group action and they have been quick to begin the organization of all types of coöpera-

THE STRUGGLES OF COOPERATION 93

tive enterprise. But in many instances the return of more prosperous conditions and the failure of some poorly planned and badly managed coöperative enterprises has led individual initiative to again take the lead with a corresponding decline in coöperation.

During the past two decades, however, American farming has reached a new stage of development and the serious problems of marketing farm products have led farmers to more persistently follow coöperative methods in order to insure themselves a fair share of the consumer's price of their product. Thousands of successful coöperative farmers' organizations are now to be found which effect the double saving of reducing expense to the farmer in the sale of his product and furnishing the product to the consumer at a lower price.

Unfortunately for agriculture, the organization of farmers into coöperative groups started to develop again at the time when the organization of labor groups was attracting much public attention and the old controversy between labor and capital engaged the public mind.

It has been to some extent a reflection of this labor and capital controversy that has led the public into believing that the farmers were organizing in the same way that labor was organizing—primarily with a view of getting a larger share in the division of profits. In most cases there has been no distinction made between coöperative selling by farmers which would enable them to farm more efficiently and market their products more cheaply and the organization of a group of wage-earners whose objective was to get a larger return for the same amount of labor or to reduce the amount of labor for a given wage.

The general fact that farmers are not interested so much whether the price of their product is high or merely moderate, so long as the relation of the market price to cost of production is such that they can make a net profit, was generally over-looked. The organization problem for farmers was altogether different from the organization of corporations by manufacturers and business men or the organization of labor groups by laborers.

Both the public and our legislatures misunderstood this situation with the result that agricultural organizations were put in the same category as labor organizations under the law and in the legislation which was designed to exempt agriculture from the handicaps of the

THE STRUGGLES OF COOPERATION 95

anti-trust act. There is no darker page in American legislative records with respect to business organization than the manner in which we have treated farm organizations, particularly the milk producers.

The record of the treatment of the Milk Producers' Association during the war is enough to point out the injustices that have been done. For some time previous to the war the milk producers in the vicinity of our larger cities had been operating successfully on a collective bargaining plan in reaching understandings with milk distributors as to the price of milk to the producer. When the war began it seemed desirable that milk production should be encouraged wherever possible, and it would have been logical to expect that every encouragement would be made to dairymen. The shortage of labor on farms had increased their difficulties and made it hard for them to operate successfully. The rising prices of feeds had added other cost handicaps.

The public attitude toward these organized dairy farmers, however, was surprisingly critical. In 1916, the milk producers near Chicago asked a modest increase in price which the distributors refused to meet. When the dairymen insisted upon the price and then began a boycott of the distributors who refused to grant a small increase, the latter were brought to grant the increased price. Immediately a storm of protest was aroused against the dairymen and there were widespread accusations of efforts to profiteer in the face of the fact that the dairyman was merely asking for a price somewhat closer to the cost of production. The press denounced the farmers and stirred all sorts of consumers' organizations to protest.

Following this popular clamor the attorney-generals of some states and district attorneys began to prepare legal proceedings against the leaders of the dairymen's organization. These authorities declared that in meeting to discuss the price of milk farmers were violating the state anti-trust laws. Indictments were returned in a few cases and this so stirred the dissatisfaction among dairymen as to bring about a critical situation. In some instances the authorities went even further, arresting and throwing into jail with scant ceremony the leaders of dairymen's organizations.

In an effort to bring order out of the turmoil, several state governors appointed milk commissions to study the situation. The commis-

THE STRUGGLES OF COÖPERATION 97

sions attempted to determine a fair price based upon cost of production plus a reasonable profit. In some instances the milk commissions were fairly successful while in other instances they failed to satisfy either producers or consumers. It was shown by almost every investigation that the cost of production of milk had risen faster than the selling price and that the farmers' demands for a better price were in most instances justified.

The attempts at prosecution, none of which were ever consummated, had a very disturbing effect upon coöperative organizations in general and stimulated the demand for a clearer definition of the law regarding coöperation so that farmers' organizations might know exactly where they stood under the law. Early in the winter of 1920, I introduced a bill in Congress to authorize collective bargaining in farm products (Senate bill No. 845). A hearing was held and representatives of thousands of farmers' organizations appeared before the Committee.

It was pointed out that this bill was not designed to change the policy of the Government but merely to explain that policy and give specific authority for coöperative efforts by farmers. It was pointed out that many states had clarified the law and put an end to indictments of legitimate farmers' organizations under the anti-trust act. All agricultural authorities agreed that the purpose of my bill was clearly in the public interest and in no sense an unfair proposal to other group organizations, nor was it class legislation. I pointed out that the right to bargain collectively was essential to farm organizations in order that they might deal with powerful groups of middlemen. Dairymen selling milk for city consumption must of necessity contract in advance, sell upon credit and make daily deliveries, and if the business was to be stabilized and the supply of milk conserved farmers must be in a position to do business the same as groups of distributors.

While some opposition had been expected, we did not look for the organized opposition which almost immediately was disclosed. The bill was held up for months in a sub-committee made up of lawyers under the claim that it was class legislation and when that argument would not stand they made the pretext that it was a threat of monopoly of the business of producing foodstuffs, or otherwise unconstitutional. No bill that had heretofore been presented to Congress

THE STRUGGLES OF COOPERATION 99

with such unanimous support found such difficulty in making progress. For some time it was difficult to trace the source of this opposition.

It soon developed, however, that the middlemen's organizations acting through Boards of Trade or other agencies were exerting pressure to prevent action by Congress. One of these organizations boldly sent a memorial to members of Congress in which it was contended that the courts and not Congress should decide whether the opposition of farmers' coöperative societies were restrictive to commerce. They also contended that the bill gave farmers the right to form an oppressive monopoly which is denied to all others.

The representatives of these opponents of farm coöperation persisted to the last in trying to deny to farmers a legal standing which the anti-trust law freely gives to any corporation, a standing which is granted to farmers' coöperative associations in practically every other civilized nation. They sought to prevent a group of farmers from doing business as freely and legally as a corporation with thousands of stockholders.

It was pointed out in the discussion which followed concerning this bill how impossible it is for the American farmers to develop an oppressive monopoly. A farm is a going concern that must keep going. It takes years to develop herds of live stock and to get a farm into operation. The farmer cannot shut down and immediately stop his cost of operation. Farms are so widely distributed and so diversified that the moment the price of a single farm product makes that product the most profitable, millions of other farmers will begin to grow the product and the increase in production brings down the excessive price. A witness before the Commission of Agricultural Inquiry pointed out that even if all of the present wheat growers of the country combined and attempted to boost prices the three or four million other farms in the country that could and would produce wheat to some extent would be tempted to go into the business and thereby check the rise in price due to the combination of the first growers. Only the possibility of organizing five or six million farmers of every class and kind would offer the danger of a monopoly.

But the most fundamental check of all against a monopoly in agriculture is the fact that no producer can determine beforehand at planting time what his crop will be at harvest. He can

THE STRUGGLES OF COÖPERATION

determine the area, the number of acres planted, but he never knows what the yield will be because that is in the hands of Providence as expressed in the Seasons and the Weather. Frequently, our years of largest acreage result in very moderate-sized crops due to lower average yield. During the war the constant stimulation of wheat-growing failed to produce a billion-bushel crop in any year except one.

The monopoly argument against farm organization has no sound basis because a successful farmers' organization must in the long run conserve the interests of the consumers who provide the farmers' market.

The demand on Congress for legislation legalizing coöperation continued for over two years but nothing was accomplished until the agricultural group compelled that body to give attention to the subject and the Capper-Volstead Act was passed. This law was hailed as a milestone in history by many and coöperative organization leaders immediately took new courage in their efforts to organize farmers. A small section of the public press, however, still clings to the fear of monopoly by farmers' organizations and their effect on the consumer's price.

For a long time the farmer has received about one-third of the consumer's dollar, the spread between the consumer's price and the farmer's price being about seventy cents on the dollar. All agree that this margin must be reduced but consumers are doing absolutely nothing about it. The entire effort has been left to the farmers and to the Government operating through various laws and public departments.

Coöperative marketing among producers is necessary to the ultimate solution of the high cost of living but coöperation among consumers is almost as desirable. I have several times proposed legislation to encourage consumer coöperation, but to the present time such organizations have been comparatively few and ineffective. In this field the consumer is being protected and benefited almost entirely by the producer while the former allows an extensive system of middlemen to expand and flourish at his expense.

The Capper-Volstead Act was the result of a most bitter struggle by those who apparently did not desire to give the farmer equal rights with corporations when he organized his business merely for mutual benefits and not for profit for the corporation itself.

The law did not seek to set up a specific form of organization for farmers since it is not the purpose of a coöperative enterprise to build up an institution which must make profit for itself. A coöperative association, if it renders effective service, keeps nothing for itself as does the corporation, but distributes its benefits to the consumer on the one hand, in the form of better distribution, better products at a more reasonable price, and to the producer on the other hand, better service in marketing his product at a reasonable cost.

The bill also gave the Secretary of Agriculture the authority to observe the development and methods of coöperative enterprises in order to determine their legitimate fields and thereby present a record of successful coöperation.

The present Secretary of Agriculture has declared that his policy will be to observe the law in a manner to encourage legitimate enterprise. Efforts will be made to assemble the facts concerning coöperation in a thorough manner not heretofore possible and this in itself will fully justify the law in that managers may be warned of possible dangers and pitfalls. The history of coöperation is crowded with stories of unfortunate experiences due to the fact that one generation has not been able to profit by the mistakes of those who preceded it. Once we have assembled the facts concerning the defects in organization which may wreck a coöperative enterprise, steady progress in this form of organization will be possible.

This legislation is only the beginning, however, as I see it, and we must proceed to lay a foundation for consumer coöperatives which will take care of the retail situation and meet the producers' coöperative organization in the wholesale market at least. I have outlined the need for a definition of coöperative enterprise as contrasted from the corporation and I believe the time will soon come when we must study the matter from the consumer's side just as thoroughly and carefully as it is being studied from the producer's point of view. While the farmer will take care of wholesale distribution, it is too much to expect that he will organize a system all the way to the consumer through the maze of retail agencies which have developed because the consumer is willing to pay for them and has encouraged them to add greatly to the cost of living.

CHAPTER X

PROTECTION FOR AGRICULTURE

THE war has opened world-wide competition for American farmers. Our magnificent resources in food production which served so effectively to aid the Allies in winning the war, though they still have the temporary outlet in Europe, must ultimately meet the competition of similar products produced with cheap labor in other countries.

The buying power of the American people is viewed with longing by those in other countries whose markets have been narrowed as a result of the war. The wealth of this country and its high consuming power per capita make it the object of the desire of traders the world over. As soon as world affairs have become better adjusted, our markets will be entered by the products of many countries whose natural resources and low standards of living enable them to produce at much lower cost per unit.

The war has also compelled us to build and develop the basis of a great merchant marine which added to the world's shipping will necessarily facilitate trade with foreign countries. One result of the world war is a wider acquaintance and an opportunity for a greater interchange of business relations.

Before the war our foreign trade in agricultural products was showing a tendency to decline in its relation to exports in general, although the total value of exports was increasing slowly.

For instance, during the period 1852-1881 agriculture provided nearly 80 per cent of all exports and in the latter year products of a total value of approximately $500,000,000 were exported. From 1882 to 1910 the percentage of exports which were agricultural declined to 50 per cent although the value increased and at times exceeded $1,000,000,000 worth of exports. From 1910 to 1914 exports increased steadily and in 1917 our agricultural exports were valued at nearly $2,000,000,000 though they constituted but 31 per cent of all exports. During the war years our exports doubled, amounting to over $4,000,000,000 worth of farm products in 1919 which constituted over half of our exports of all sorts.

At the same time, from 1881 to 1918 agri-

PROTECTION FOR AGRICULTURE 107

cultural imports had amounted to close to 50 per cent of our total imports. The question is asked: Can we expect that agriculture will continue to provide the same proportion of foreign trade and to furnish exports in values that will meet the imports and thereby balance the books with respect to farm products?

The world as a whole is impoverished greatly by the war and the efforts by all consumers must be to buy in the cheapest markets. Ocean transportation is relatively cheap; consequently the flow of products from one country to another must be comparatively free. Nearly every nation except the United States is in great need of increasing its national earnings by a favorable balance of trade. Consequently they will sell everything that can be sold to the United States. What must be done to protect American agriculture and adequately provide for its future development at a rate to correspond with our national growth? The answer is to be found in a carefully adjusted tariff protection followed persistently as a national policy.

Under the policy of protection we have built up a great industrial nation and the same protection cannot now be withheld from agriculture if we would preserve the balance between

108   THE AGRICULTURAL BLOC

industrial and agricultural growth. The absence of adequate protection for farm products during the period of the last administration when the Underwood Tariff Act was in force was offset by the unusual demand which arose for food due to the world war. Our agricultural exports had begun to decline. Our city population was increasing; new lands were no longer available and more intensive farming was becoming necessary. The inevitable result was that costs began to rise and as costs rose the importation of agricultural products competing with our own increased. The war intervened and changed the direction of the trend for a time.

For years the tariff was of comparatively little value or importance to the American farmer except in the case of a few products like wool and sugar. In most cases farmers received but little benefit from the tariff. Our southern states, for instance, were almost solidly opposed to protection because their main product, cotton, met with no competition from abroad. But now, at the end of the war, we find an insistent demand for tariff legislation from the south which will protect the farmers of that region from the competition of products appearing in our market which were heretofore unknown. These have come chiefly in the form of vegetable oils which compete with the by-product of cotton-seed.

The principle of protection has been supported because we have believed it for the best interest of the country to develop our own rich resources and to employ our own labor and business ingenuity. We have sought to diversify our life and afford our growing population the widest range of employment. The agricultural states have supported this principle and aided in sheltering our industries from foreign competition. They have believed in protecting the home markets for the manufacturer. We have now arrived at the problem of protecting the home market for our agricultural products as well.

A properly adjusted tariff with relation to trade in agricultural products is now a necessity more than ever before. This is best emphasized by reviewing what has occurred with respect to some of our leading agricultural industries.

The sheep industry is an outstanding example of the influence of protection. When, in 1913, raw wool was put upon the free list, there were over 51,000,000 sheep reported on the farms in the United States. In round numbers we had about 37,000,000 sheep of shearing age. In 1921, the number of sheep had declined to 45,000,000 head and the depression in the industry was acute. During the same period, the cost of raising sheep increased nearly 200 per cent and the costs for wool alone were several times the costs ten years later. In the meantime the sheep and wool industry in countries that compete extensively with the United States changed but little. Last year while our own wool clip was still in storage we were importing ten to twenty million pounds a month even while our wool-growers were unable to pay their bills or get credit with which to buy feed to winter their sheep.

Why should we continue purchasing mutton and wool from farms thousands of miles away when our own producers are suffering because of a limited market? There is only one answer: In order to restore the balance, there must be protection for the industry which will place American producers on a fair basis to compete with the imported product.

The stabilizing value of a protective tariff must not be overlooked. Our producers cannot continue to produce sheep and wool in the face

PROTECTION FOR AGRICULTURE

of the constant danger of heavy imports at any time when the price rises to profitable levels. Bankers and other financiers are not encouraged to aid farmers in establishing economical production if the market prospects are indefinite. The war demonstrated how important the wool supply is in such events through the wild scramble for wool which occurred early in the war period. As a measure of national protection a sheep industry that will produce a considerable proportion of our requirements is good public policy.

The competition of foreign cattle and hog producers has not yet become so serious as with the sheep industry, but our surplus of these kinds of meat is decreasing and imports offer considerable menace to the market in the future. Our exports of pork products were so stimulated and increased during the war as to detract attention from the danger of foreign competition.

It is generally true that an imported quantity of the farm product exercises a depressing influence on the price to American producers out of all proportion to the amount of the imports in relation to American production. The price effect of imports is even greater than the relation of imports to our own surplus. The measure of protection to be given must be fixed on a basis of stabilizing a price rather than merely for protection from imports.

Foreign competition is less serious to the wheat-grower and while admitting that the value of a tariff on wheat is largely psychological it is none the less important. So long as our large export surplus of wheat continues to be demanded by Europe, protection will be of less importance than when the wheat production of the world is restored more nearly to normal. But sooner or later we will suffer from the competition of countries where the cost of production is lower than it is in the United States.

An outstanding example of a new form of foreign competition which did not exist before the war is in the case of vegetable oils. For a decade our cotton-seed oil has been the chief oil used in food products. Suddenly there began to be imported quantities of vegetable oils from the Orient which competed directly with cotton-seed oil. The flow of these oils was directed to America by the secession of ocean shipping to Europe due to the submarine campaign and there developed within the United States ex-

PROTECTION FOR AGRICULTURE 113

tensive industries based on the use of these oils. American peanut growers met a new competition from the Orient, where this crop is produced with very cheap labor, and even the producers of dairy products felt the influence of this new import movement of a product not heretofore found in our markets.

The dairy industry is growing rapidly in several parts of the world, to an extent that may develop serious competition for American dairymen if they are not adequately protected. It is becoming more generally known that milk is one of our most essential foods to maintain the health and virility of our people. We need not emphasize the great importance of the dairy industry. Not only is the dairy cow a very efficient food-producing machine but she is almost a necessity to a permanent system of farming because of her value in maintaining fertility and in converting rough stock feeds into the best of human foods. The costs of producing dairy products have been steadily increasing, not the least of which is the increase in the cost of labor in handling the dairy herd and its products.

This review touches only a few of our more important agricultural products, but its importance is shown by the fact that the value of our farm production in a single year is more important to us than the entire indebtedness of foreign nations to this country. There are those who complain that our tariffs must not restrict our trade with Europe and thereby handicap Europe in its efforts to repay debts to this country. American agriculture, however, is of so much more importance to us than these foreign debts that it would be better that these debts were never paid than to injure our agricultural industry.

The Emergency Tariff Act which was primarily designed to protect agriculture had a far greater psychological effect than is generally realized. Though it is difficult to show just how great was the effect of this act on prices of farm products, it is not difficult to point out that a return of better conditions was to be noted shortly following the enactment of that act.

The United States is one of the world's greatest consuming markets and our farm people constitute roughly one-third of our consumers. Any protection, therefore, that tends to conserve the buying power of our agricultural people is conserving the market for industry and commerce. The admission of cheap imports from foreign countries free of duty may be a temporary benefit to consumers, but the price to the nation is a serious injury to both agricultural and other industries and a more far-reaching effect in depressing our standards of living.

We need a new method of making a tariff. The old system of collecting over 500 men from all parts of the country, many wholly unfamiliar with world-wide economic conditions, and expecting them to draft a tariff bill which will stand the test of constantly changing conditions is wholly out of date. Add to this the complications of inherited ideas concerning the tariff by the descendants of partisan leaders who regarded the tariff purely as a political foot-ball, and you have a picture of the difficulties of making a tariff in Congress as a whole.

A great mass of contradictory testimony, absolutely impossible to digest or reconcile, has been assembled. Much of the testimony has been presented by interested parties rather than unbiased experts. Then, finally, the entire bill is subjected to interminable and heated debate and is passed as a result of compromise largely to get rid of the nuisance. Since the tariff bill was first written in 1921, conditions in trade have entirely changed and other changes may occur over-night in foreign countries which our old-style tariff cannot be adapted to meet.

A scientific tariff plan has been talked about for generations and in sober moments nearly every thoughtful man admits that the tariff is an economic and not a political question. We have followed the adoption of the most up-to-date methods in every other field of government endeavor by dragging out the old rusty machine and endeavoring to oil it up in this time of the world's chaotic economic condition. Never before has there been such a general realization that we must begin now to develop our tariff-making machinery in keeping with the requirements of our foreign relations, which are certain to grow more complicated as trading with foreign countries increases.

The bill introduced by Senator Frelinghuysen, of New Jersey, was a great step in the direction of the desired end. If it is possible to go further and set up machinery which will provide for tariff adjustment as conditions change rather than as political control changes, we shall have gone forward toward giving justice to all who are interested.

The Agricultural Bloc as a unit has not discussed the tariff because of its present political nature, but there is no question that there is a growing sentiment that the only just tariff to agriculture must be settled on the basis of economic justice. Conditions in world-production of agricultural products may change very quickly and make it desirable that there be immediate tariff adjustment upon an economic basis. Foreign countries have learned through experience to make their tariff adjustable and have provided quick-acting legislative machinery to do the job. Even in free-trade England, where tariffs are again beginning to appear, Parliament merely lays down the general lines of protection and does not attempt to fix rates or details, but leaves that to a competent Board of Experts.

CHAPTER XI

THE PUBLIC ATTITUDE TOWARD THE BLOC

WHEN the organization of the Agricultural Bloc first received public notice in Washington there almost immediately developed a spirit of opposition on the part of the representatives and the press of our large cities. The New York newspapers were promptly critical and did not stop by merely announcing the organization of the group, but went further to state what they assumed were the objects and motives of the group and to discuss them in detail.

The New York *Journal of Commerce* said:

> "The whole program—or certainly the major portion of it—is based upon false premises. The congressional theory seems to be that farmers are being exploited by other interests, no inconsiderable part of which live as parasites upon the agricultural communities. The logical procedure is therefore (they reason) to enact legislation which will free the farmers from this burden. Unfortunately, however, many are not willing to stop even there, but on the contrary, desire special government favor or a subsidy for agriculture."

Such a point of view was passed on to business men throughout our largest cities to such an extent that it was almost immediately reflected back to Congress into the minds of representatives who had not taken the trouble to study the facts themselves, but merely reflected the opinions of their constituents.

The criticism largely took the form of editorial ridicule until there came the first real test of strength of the Bloc in July, when the Senate leaders endeavored to adjourn the Senate without giving attention to pending farm measures. Then a new fear seized upon the city press and they devoted much space to the danger to organized political party action from this bi-partisan movement. The Philadelphia *Public Ledger* said:

"The significance in this particular blow lies in the overriding of party regularity by both Democrats and Republicans of the 'Farm Bloc' that has risen to power and pride of place in the last two months. This is a class, an industrial and geographical group, that has broken through party lines and ignored them. At least one million of the hitherto voiceless American farmers have become vocal with startling suddenness. . . . The farm legislative program is loaded with certain near-radical proposals, nationalization possibilities, and federal regulation schemes. . . . In a straight out and out fight the conservatives lost and the Farm Bloc won an important encounter that will bring on a pitched battle."

A few days later the same newspaper forecast a determined fight between the Farm Bloc and the rest of the Senate and this time stated that 2,000,000 organized farmers were behind the movement. It is to be noted that this section of the press at least was discovering the far-reaching support which was behind the agricultural Senators.

The misunderstanding of those who are unfamiliar with the agricultural situation continued to grow, and many dangers were cited for which the Bloc was in no sense responsible. The growing tendency towards group organization drew criticisms from the press and the menace of this group movement spreading to other interests was outlined. It is notable that the press appeared to have forgotten its discussion of organized lobbies of various sorts of a year or two before and the almost constant agitation of the control exercised by Wall Street, which we observed prior to the war.

Then other situations and developments in Congress entirely outside the scope of the agricultural group began to be attributed to the Bloc by popular statements. The opposition to certain parts of the tax bill as well as the support which certain measures received were attributed to the influence of this group. Had its members been disposed to do so, they might easily have taken credit for many things which they were in no wise responsible for. However, the accomplishments of the year were relatively few outside of matters having a primary interest to agriculture. In one instance, a small group of Senators chanced to meet at my home for an evening conference on a matter not of particular agricultural importance. The reporters immediately assumed that this was a meeting of the Bloc, and proceeded to note the actions of all those who were present and attribute them to an imaginary agreement reached in this meeting.

After a time the more discerning newspaper writers seeking for a new angle from which to treat what was now becoming a familiar story, began to discover that there was nothing so very dangerous coming out of the Bloc's activities and that its program on the whole was inclined to be constructive. One writer in the *Annalist* of New York said:

"As yet there is nothing ultra-radical about the movement, but it is distinctly against the acceptance of a stand-pat administration program, and the plain facts up to this time supply the proof that more than one monkey wrench

122 THE AGRICULTURAL BLOC

has been thrown into the machinery which many predicted would move smoothly enough when the administration went into power."

The *Public Ledger* said:

"The hoof marks of the 'Farm Bloc' decorate many sectors of the administration profile. The embattled farmers from the Corn Belt and the Cow Country have mutilated the Harding program and made the administration eat much dirt on the tax, tariff, railway and other pet White House measures."

It is interesting to compare this statement with the summaries of the administration's accomplishments made from time to time by administration leaders in the House, which show that many of the measures put through primarily because of the support of the Bloc were also looked upon as those of greatest importance by the administration.

The press found a sensational topic in the controversy between the Bloc and the President and missed no opportunity to build up public interest in the discussion on this subject. When Senator Kenyon was first offered a position as Federal Judge in Iowa there was considerable glee in certain quarters over the prospect that the Bloc might soon lose its leader. These writers did not take the trouble to inquire into

PUBLIC ATTITUDE TOWARD BLOC 123

Senator Kenyon's personal interests in the Bench, or to discover what was known by many of his friends, that he had not settled down in the Senate for life.

The conclusion seemed quite general in the minds of editorial writers that the Bloc's objective was class interest and nothing else. Some of the class representative journals expressed this view with satisfaction, being pleased that their opponents, the other special interests, had a new and powerful opponent in the farmers' organization.

The most vigorous opposition from eastern cities hinged around the discussion of representation for agriculture on the Federal Reserve Board. The New York *Times* said:

"No other interest demands that the national banking policy shall be molded by class preference."

No one had suggested that the addition of one representative of agriculture to a reserve board consisting of seven able men could be expected to mold the national banking policy entirely for agriculture. The entire movement was merely to give agriculture a fair representation along with other of our great industries. Certainly the biggest business in the country, basic to all other industries, with a greater capital value, was not asking to dictate Federal policies when it asked to be represented by one man on a board of seven.

Not all of the press, however, was unsympathetic or lacking in discernment. Some of the newspapers whose editorial writers had given more attention to studying the agricultural situation were frank to point out the facts as they saw them. The Washington *Herald* said:

"In turning to a fair examination of what the Bloc has done with its power, it must be said that so far as these measures are concerned, which they initiated or advocated, little can be said in the nature of criticism even by those who are mostly disposed to deplore the underlying principle of group action in politics. . . . One of the most striking events in the present Congress was frequently but erroneously attributed to the leadership of the Farm Bloc, that was the forcing of the retention of high sur-taxes on large incomes."

The New York *Commercial* said:

"If the so-called Farmers' Bloc can develop legislation that will reach the fundamentals it will have not only the sympathy but the support of the entire country. . . . If on specific problems they will devote themselves to the development of better marketing facilities . . . they would be getting down to the root of their own troubles and those of the nation at large."

PUBLIC ATTITUDE TOWARD BLOC 125

Another common mistake in the press was to attribute activity within the Bloc to Senators who were never identified with it. Almost from the outset the membership in this group was definitely fixed and there were but few changes.

The agricultural press, being thoroughly familiar with the farm situation, was able to present a much more accurate picture to the farmers of the country than city folks were able to get from the metropolitan newspapers. The changes that were constantly taking place were observed by farmers and they understood the logic of the shifting in legislation having to do with farm needs. This section of the press pointed out with considerable amusement how the business groups had suddenly developed a great antipathy to the grouping of legislators along economic lines instead of party lines and their evident distress over the tendency to deplore representation by specific industrial groups.

The attitude of certain opposing interests was clearly reflected in the trade press, at times, when they would bitterly assail the Bloc as being an opponent for selfish class interests of the best interest of the nation. These papers furnished an excellent index to what manufacturers were saying among themselves when agricultural measures began to get the deserved attention.

Had any other single industry experienced a similar decline in the value of its products while it was still in the hands of the producers that farmers experienced during 1921, there would have undoubtedly been a panic of tremendous proportions. When an $80,000,000,000 industry which was supplying the nations over $20,000,000,000 of new wealth annually and 50 per cent of its bank deposits was flat on its back asking for help, conditions would seem to be ripe for a general panic. When one-fourth of the 6,000,000 farmers of the country were in a position practically bankrupt, it is a serious matter. Some said that because these were farm tenants it didn't matter so much. But the farm tenant is in no different position from the small merchant who is renting the building in which he operates his store and we know that had one-fourth of our merchants been in bankrupt condition panic would have been a mild term to apply to the situation.

Now that it is a matter of history, we hope, the tremendous decline in farm values can be accurately measured. The Department of Agriculture in its report on the value of farm products, crops and live stock for the last three years gives the following figures:

	1919	1920	1921
Value of crops.	$15,422,000,000	$10,909,000,000	$7,027,000,000
Value of animal products	8,360,000,000	7,354,000,000	5,338,000,000
Total...	$23,782,000,000	$18,263,000,000	$12,365,000,000

Such a report was not apparent to the public at the time the depression was going on, but farmers knew that it was real and the wonder is that they were able to stand fast during this experience and emerge with a determination to renew planting for another year and grow another crop so essential to the nation's business.

The story of the tens of thousands of farmers who did not escape bankruptcy will never be known to the city consumers who have dined comfortably on the products of the labors of these farmers. Scattered through the far reaches of our country, one by one, these farmers saw their savings of a lifetime disappear and they could only pack their small stock of tools and move on to another piece of land to begin their struggle over again. Those of us who represented purely agricultural states in

Washington found our mail-bags filled with letters from strong men who were down, but not out. Only one who has lived among farmers could read between the lines the unwritten story of the privations and disappointments and actual tragedy which these letters carried. And yet, the great mass of our people in the cities, living in comparative comfort, read the misleading editorials in many newspapers and believed that the Agricultural Bloc in Congress was based upon a selfish class interest.

More recently the suggestion has constantly been repeated that there was soon to rise a new party, some sort of an agrarian movement which would sweep the farmers of the country into its ranks. It has been argued that the farmers of the country were becoming politically self-conscious and were dazed by the knowledge of their power when organized.

That there are no grounds for such an outcome to the situation can be demonstrated by examining the political balance of the groups that have been organized in Congress. The political managers have been obliged to revise their tactics, however, since there are two outstanding features of the Bloc's activities which have been disconcerting to many of them. The first is the open and above-board methods that have been followed in this contest. The fight has been entirely in the open where all could see what was being done and the advice of those who were recognized as leaders was constantly followed. In the second place, old traditions and sectional hostility, particularly between the farmers of the north and south, have been swept aside in the interest of the national good.

It was not until early in 1922 that a change in the disposition of the public press began to be evident. Since then, there have been quite a few leading newspapers and journals which have pointed out the similarity between the program of the Bloc and the platforms of both leading parties and particularly the charges which the President has given to Congress at various times.

To those of us who have observed this movement closely from its inception, it is clear that this is not primarily evidence that the farmer is depending more upon legislative remedies for economic difficulties, but rather that he insists that his representatives in Congress shall use every possible means to secure their reasonable division of wealth between the agricultural producer and the worker in other industries and commerce. He sees in this movement an opportunity to assure himself representation which will really represent. Congress has always been largely composed of lawyers. Even the large agricultural states have been represented by delegations most of whom were lawyers. The desire to put farmers in Congress has at times been expressed as a means to get better representation and that would undoubtedly help a great deal, but the more important thing is to have representatives who are studying the needs of agriculture, constantly advising with farmers' representatives, in close touch with farmers themselves, and who have an appreciation of the fundamental importance of agriculture in national prosperity.

CHAPTER XII

THE FARMERS' PROGRAM

POLITICAL movements among farmers have generally been viewed with alarm if not openly ridiculed by the representatives of the population's centers during recent years. Otherwise careful students of current history have been prone to misjudge farm movements and to question the motives and purposes of the leaders. Those who have lived among farmers and understand their way of thinking and observed the causes for political discontent are seldom so disturbed as are the politicians from urban centers.

The rise and growth in influence of the Non-Partisian League in North Dakota revived the agitation in the eastern press and brought on an epidemic of uncalled-for comment and criticism about American farmers. The result was that the growth of farm organizations which has been quite general since the war has more seriously disturbed some people than any other political movement or even the over-turn of affairs by the world war. Some of the very provincial self-styled statesmen of the East worked up a greater fear of the wave of unrest which was sweeping the country-side than they had experienced for a long time.

We were faced with a constant discussion of the spread of Bolshevism, Red doctrines and Socialism among our farm people. The nation was warned time and again that its foundations were being undermined by the very people upon whom we had depended for standards of liberty. Minor incidents such as occur in some communities every day were given wide publicity as evidence that the farmers of the country were becoming dangerously radical. Sensational writers sought for material to feed this popular demand and were able to supply instances enough to keep the agitation going. Business men from eastern cities who had apparently assumed that the old reliable farmer was still thinking in the terms of 1890 and had not been keeping abreast of national progress became decidedly nervous and jumpy because of this imaginary danger of farm radicalism.

There were a few professional agitators who lived by leading insurgent movements among groups of farmers who had a lively day of popularity. Cool-headed farmers knew that most of the accusations were groundless and would pass with the change in the public temper, so they gave but little thought to the situation.

As it happened, the organization of powerful national farmers' associations with the purpose to develop business and educational improvements was taking place at the same time. The nervous city press was unable to distinguish between a conservative business farm organization and a radical group which met occasionally to relieve its feelings in a protest meeting. The result was that farm organizations as a whole began to seriously disturb quite a lot of the city people and not a few of the members of legislatures and Congress. Without taking the trouble to secure the facts, they jumped to the conclusion that the farm organization movement was headed toward socialism or something of the sort and proceeded to object to anything that emanated from an organized farm group.

The result of this situation was that when the Agricultural Group began to meet in Washington there was a perfect panic in the editorial minds of some of our large city publications with the fear that this farmers' revolutionary movement was reaching into the seat of government and menacing the heart of the nation. This attitude would have been highly amusing if it had not been so misleading to the great mass of city people who were not in a position to know the facts yet whose interests were being protected by the stability of the very farm people who were being so seriously misjudged.

Numerous farmers' meetings throughout the country laid out programs of policy and action which should have been sufficient to convince any one that the farmers' national program was perfectly in keeping with the best interests of the nation. One national farmers' convention after another would meet and endorse a constructive program. There was a decided agreement between these organizations so that one might conclude that their resolutions were written by the same author. There was an absence of radical proposals except in a very few instances which received far more publicity than their importance deserved.

It remained for the President at the suggestion of the Secretary of Agriculture, Henry C. Wallace, to call a conference which would show the temper of the people of the farms in unmistakable fashion. The National Agricultural

THE FARMERS' PROGRAM 135

Conference held in Washington January 23-27, 1922, brought together representatives of all important groups of farmers of various sections, together with those engaged in the business of handling farm products, thereby providing a cross-section of agricultural thought such as we have never had in our history heretofore. Among the 336 delegates there were 275 farmers and 25 farm women. They were brought together to outline the present needs of agriculture, to propose a future program and to give voice to the real sentiment and desires of farm people.

When the conference was opened by President Harding, he presented an address containing the strongest argument for agricultural advancement which has been heard from a chief executive since the time of Lincoln. It is worth attention at this point to review some of the points which our President made in that address concerning the national program for agriculture. A series of quotations will suffice. He said:

"Concerning the grim reality of the present crisis in agriculture, there can be no differences of opinion among informed people. . . .

"Now, in his hour of disaster, consequent to the reaction

from the feverish conditions of war, he comes to us asking that he be given support and assistance which shall testify our appreciation of his service. . . .

"Agriculture is the oldest and most elemental of industries. . . . It is the first industry to which society makes appeal in every period of distress and difficulty. . . ."

He spoke at length of the needs for better farm credit, for legislation which would enable the farmers to do coöperative marketing, for extended marketing information, and urged the need of farm organization. He said:

"The farmer does not demand special consideration to the disadvantage of any other class; he asks only for that consideration which shall place his vital industry on a parity of opportunity with others and enable it to serve the broadest interests. . . .

"There must be a new conception of the farmer's place in our social and economic scheme. . . ."

At this time, the President, like many others, did not quite understand the purpose of the Agricultural Bloc in Congress and did not appear to realize that it had aided in the passage of several bills which would be regarded as the biggest accomplishments of his administration.

The Agricultural Conference heard reports on the situation in agriculture in various parts of the country and in various branches of re-

THE FARMERS' PROGRAM 137

lated industries. It then divided into committees through which to put the thought of American farmers regarding the present and future situation into the form of a definite program.

There had been considerable fear that a radical program would be outlined with propositions for price-fixing and special attention to agriculture at the expense of other industries and groups of people. But any one who will carefully examine the reports of these committees will be unable to disclose anything but constructive, sound policy which would reflect credit upon any group of people.

Early in its resolutions was one commending and approving the action of the Agricultural Bloc in the following terms: "We commend and approve the action of those members of the House and Senate comprising the Agricultural Bloc who, regardless of party, so early saw the emergency and have so constantly supported a constructive program for the improvement of agriculture and the betterment of rural life."

The legislative recommendations growing out of the conference included the endorsement of practically every subject on the program of the Agricultural Bloc as well as many other subjects not considered by the Bloc. Among those specifically endorsed were bills to provide for short-time agricultural credit; an agricultural representative on the Federal Reserve Board; repeal of section 15a of Interstate Commerce Act; the operation of the Muscle Shoals project; enactment of coöperation legislation and credit based on warehouse certificates.

The Conference served to distribute throughout the country a more accurate knowledge of the objective of the Agricultural Bloc in Congress and the manner in which it was seeking to aid agriculture through legislative methods. It served the double purpose of convincing Senators and Representatives in Congress of the soundness of their program; its agreement with the needs and desires of farmers, as well as to inform the farmers of the country of the situation in Washington and the unreasonable opposition which had been encountered by many bills which were in no sense selfishly proposed by the farmers' friends.

Dating from this time, January, 1922, there was to be noted a steady change of attitude concerning the farmers' organized movement. Criticism in the city press began to subside. The comments upon the action of the Agricultural Bloc in Congress became less caustic and

THE FARMERS' PROGRAM 139

more sympathetic. While the opposition from certain quarters grew more determined the support of the rank and file of farmers had been clinched by the contact made at the Agricultural Conference and by the fact that it was disclosed that the farmers' program and the program of the Agricultural Bloc were identical. It could not well have been otherwise because of the primary fact that the Bloc program was based upon a study of farm conditions and intimate conference with farm leaders.

CHAPTER XIII

THE PROGRAM OF THE BLOC

At the first meeting of the Agricultural Bloc, Senator Kenyon in his introductory remarks reviewed the generally deplorable condition in agriculture and pointed out the necessity for action by Congress on relief measures. He suggested that by bringing together Senators from the leading agricultural sections of the middle west and south it might be possible to bring to bear sufficient coöperative action in the Senate to put before that body some of the pending measures.

The farmers' desires were then reviewed by speakers who emphasized the general interest among farmers in proposals for better financing for farm marketing, personal credits, the privilege of coöperative marketing, the readjustment of freight rates, packer control and the regulation of future trading in grain. In fact the entire list of agricultural measures of importance was discussed.

The suggestion came from many sources that

there had been plenty of discussion and what was needed was prompt action. It was apparent that such action could not be secured unless the members of Congress could agree to take up a few measures at a time and push them steadily toward final passage. The purpose of this first meeting of the agricultural group of Senators was to reach some sort of an understanding concerning a program of action and to agree as to the measures of first importance in the emergency.

The need for credit for farmers was clearly shown to be of paramount importance. It was strongly urged that the Federal Reserve Act be amended to provide the needed emergency credit for farm financing and if possible to secure for agriculture the same advantages under the Act that were afforded other industries. The conclusion of this meeting was that there were four groups of bills which should be considered of primary importance and sub-committees were appointed as follows: 1. Amendments to the Federal Reserve Act; 2. Plans of Commodity Financing; 3 Transportation; and 4. Other Miscellaneous Measures.

The discussions between the members of the Agricultural Bloc continued from time to time as these committees surveyed the situation. At that time there was no definite plan of procedure, but it was perfectly clear to the Senators familiar with the agricultural situation that there were certain outstanding needs for farming that needed first attention.

Early in June, another meeting of the Bloc was held and committee reports were heard. Those who had been studying ways and means of extending more and proper credit to agriculture urged that the Federal Reserve Act be amended so that credits might be extended to twelve months on live stock and farm products instead of six months as provided in the law. It was made clear that the proposed amendments should be carefully studied and discussed fully with representatives of the Treasury Department and others before any action was taken. In all of these early meetings the disposition of the members of the group was to avail themselves of every possible source of information and advice before urging a piece of legislation before the Senate.

In the meantime, several of the bills which had been pending before the House and Senate had come up for consideration and the Bloc coöperated in supporting these measures to ad-

THE PROGRAM OF THE BLOC 143

vance them on the calendar. In fact it was hardly necessary for the group to make a more definite program than merely to unite on nonpartisan lines in support of agricultural measures when they came before the Senate. The record of action in the Senate during the months following the organization of the Bloc, shows the results accomplished.

In reviewing the accomplishments of the session in August at its close, the Republican leader in the House pointed out that few Congresses in American history had made a better record of progress through the hot summer than this, the opening session of the new administration. In the long list of measures that had been passed, there were represented more measures of interest to agriculture than are to be found in the action of any similar session in recent years. Those bills which were of particular interest to farmers were the following:

The Emergency Tariff, which was principally on agricultural products, was extended.

The Fordney Tariff Bill, containing higher protective rates for farm products, had been passed by the House.

A Treasury deposit of $25,000,000 for the Federal Farm Loan Board was provided.

The Grain Futures Bill, to prevent gambling in grain futures, a subject long discussed, had been passed.

The Packer's Control Act, which had been pressed for ten years, was enacted, bringing to a close a long-drawn-out controversy on this subject.

The McNary Bill amending the War Finance Corporation Act to provide relief for producers of agricultural products was passed.

The end of this session of Congress found the Agricultural Bloc in the Senate working steadily along the lines laid out at the beginning with no change or deviation except as had been found advisable as more careful study of the projects developed. Early in the Fall the Bloc took up the program again with renewed energy and placed first on its list for favorable action the enactment of the Capper-Volstead Coöperative Marketing Bill which had long been pending in the Senate and had received but limited support.

It was decided that the bill providing for representation for agriculture on the Federal Reserve Board should be supported and the bill so drafted as to secure the appointment of an actual farmer upon that Board. It was also

THE PROGRAM OF THE BLOC 145

decided to work for bills having to do with the Federal Farm Loan Board to extend its scope of usefulness.

It is of interest to mention here an incident illustrating the constructive methods of the Bloc. The Federal Farm Loan Board had long been criticised for a lack of vigorous activity in pushing its loans. Legislation was suggested looking to the reorganization of this Board in order to make it function more satisfactorily. But, rather than push legislation alone, the Bloc appointed a committee to call upon the Federal Farm Loan Board to ask for more vigorous action in meeting farmers' applications for loans and promoting the sale of farm loan bonds.

The results of this campaign and the soundness of the judgment of those urging a more rapid sale of bonds was indicated by the fact that succeeding issues met with prompt sale when offered and there was almost immediately a speeding up of farm loan activity. Legislative action was not taken immediately and the changes in attitude of the Farm Loan Board may be attributed largely to the spur instigated by the Agricultural Bloc.

By this time, the Bloc had been in action for

some six months and its determination and effectiveness was coming to be well understood. It was discovered that suggestions in various quarters would accomplish the result almost as quickly as action in the Senate. When it was realized that the Bloc had organized for action and not merely for agitation, its influence increased.

From the very first meeting the Agricultural Bloc was favored with the constructive advice of leaders who were familiar with the agricultural situation as well as numbers of other men prominent in national affairs, business and industry, who realized that the improvement of agricultural conditions was the first step toward the revival of prosperity.

In this group must be named Secretary of Agriculture Wallace, Secretary of Commerce Hoover, James R. Howard, President of the American Farm Bureau Federation; Charles S. Barrett, President of the Farmers' Union; S. J. Lowell, Master of the National Grange; Fred H. Bixby, President of the American National Live Stock Association; Gifford Pinchot, Barney Baruch, Eugene Meyer, Smith Brookhart, Aaron Sapiro, and Thomas A. Edison.

THE PROGRAM OF THE BLOC 147

Most of these men appeared before meetings of the Bloc upon invitation to discuss specific measures. Many of them conferred with members frequently in respect to legislation.

Supplementing the advice of these leaders the members of the Bloc were in constant touch with the representatives of farm organizations located at Washington, prominent among whom may be named Gray Silver, of the American Farm Bureau Federation; Dr. T. C. Atkeson, of the National Grange; Charles A. Lyman, of the National Board of Farm Organizations; Charles Holman, of the National Milk Producers' Federation and many others. At the very outset, Senator Kenyon, the Bloc leader, requested the support and advice of all who were in close touch with farmers' needs.

A long list of proposals of various sorts by various individuals in Congress were attributed to the Bloc which in reality were never discussed or considered by the group at all. Any agricultural measure that received favorable comment was likely to be called a product of the Bloc even though it was proposed by some one in no wise connected with the group. In many instances group action which bore no relation whatever to the program of the agricultural group was pointed out as a part of its program in the press. From the very outset there were certain questions which the Agricultural Bloc regarded as distinctly partisan and outside the scope of a bi-partisan organization.

One instance was the agricultural tariff. Though there has come to be a surprising agreement between representatives of the South and West on the agricultural tariff, it has never been a part of the Bloc's program. Certain other distinctly administration measures have been regarded as foreign to the central purpose of the Bloc and have never been introduced in its discussion. It was inevitable that group action when it had once been demonstrated to be as effective for agriculture as it has been in the past for business, industrial and labor groups would be adopted by those supporting other measures. The "Farm Tariff Bloc," so-called, was from the outset a different group though necessarily including many of the Senators who were members of the larger group.

The critics of the agricultural movement have failed to realize that there is not an entire agreement between farmers of various regions because in many instances their best interests lies in opposite directions. There are numer-

THE PROGRAM OF THE BLOC 149

ous subjects upon which farmers from different parts of the country do not agree, not because they differ in political faith but because of natural competition. Such questions were omitted from the Bloc's program. One illustration of such a case is Reclamation Measures very vital to the people in certain western states but opposed by farmers in the older established regions who do not feel that the development of new lands is wise at this time in view of the fact that agriculture as a whole is not sufficiently profitable. Many eastern farmers will agree with those from the west regarding the need for better rural credit, improved marketing and lower freight rates, but they disagree when it comes to a question of federal aid in the reclamation of new land which will bring products into competition with those from older farms.

Early in 1922 the Agricultural Bloc reviewed its program, counted up its accomplishments, took stock of the situation and made out a plan of action for the new session of Congress to begin in the spring. Senator Kenyon resigned from the Senate to accept the judgeship and the writer was chosen chairman to succeed him.

The regular practice of hearing the farmers' representatives describe the farmers' needs and desires was continued. At one of the meetings of the Senate group this spring there were present representatives of the leading farm organizations who spoke for the national headquarters of practically all the organized farmers in the United States. After hearing these men together with other experts on the subject of rural credits a special committee was appointed to make further studies before a bill should be selected to have the support of the entire group.

This new program included the following measures:

Farm credits legislation—The enactment of a bill or a combination measure which would meet the requirements so clearly set forth by the Joint Commission of Agricultural Inquiry:

Extending the War Finance Corporation until the need for special credit had passed or new rural credit legislation had been enacted:

The removal of the $10,000 limit on Federal Land Bank loans in order to enable loans to be made in states where the average value of farms and buildings was more than twice the old limit. The average value of farms of the entire United States for the 1920 census was $12,084, compared to $6,444 in 1910. The limit

THE PROGRAM OF THE BLOC 151

of placing first mortgages at $10,000 made it impossible to cover even 50 per cent of the valuation on many farms:

The Truth in Fabrics bill, designed to assure the elimination of fraud in the sale of woolen fabrics. While this legislation had moved slowly, due to the opposition of the powerful manufacturing interests and the difficulty of working out an effective method of administration, it was none the less vigorously supported by the members of the Bloc:

The Filled Milk legislation, designed to prevent the manufacture and sale of products as milk containing vegetable fats, was listed for attention.

The Muscle Shoals development plan for the purpose of producing fertilizers was approved and listed among those measures that should have prompt attention.

The policy of the Bloc to specialize in the study of one particular measure at a time, go thoroughly into its various angles, and to work out a well organized plan of procedure was still kept in the fore-ground. The Agricultural Bloc's program has never been extensive at any particular time since its whole purpose was to concentrate upon matters of current importance and secure action without undue delay.

When from time to time the press has appeared to note a lull in the activities of the group, it has merely been assembling the facts with which to forge ahead with the new measure. While there have been many who have prophesied that this form of group action would soon pass into history along with other political incidents, there are yet so many unsolved questions that require consideration that there is plenty to do for some time to come.

CHAPTER XIV

THE RECORD OF THE BLOC IN CONGRESS

ACTION on agricultural matters in the Senate dates from the first meeting of the group of Senators in May, 1921. The agricultural depression had been thoroughly discussed, both in hearings and on the floor of the Senate, but little had been accomplished except the preparation for the passage of the Emergency tariff, generally called the "Agricultural" tariff bill. There had been much talk and little action since the end of the war.

While tariff questions were never regularly considered a part of the Bloc's program, there is no doubt that the attention which agriculture has received in the preparation of the tariff has been to some extent due to the presence of the Bloc. A group of Senators worked together on the tariff and this group was sometimes called the Agricultural Bloc, but this was not correct.

Shortly following the time when the Bloc began to consider its program of action the Emergency Tariff bill was passed. Another measure which was not a Bloc matter but supported by several members was the authorization by resolution of the Joint Commission of Agricultural Inquiry. The presence of the Bloc undoubtedly speeded up the appointment of this Commission, which got into action in June and included in its membership several members who were identified with the Agricultural Bloc.

Among the bills which it was first proposed to push was the amendment to the Farm Loan Act providing for an increase of working capital of $25,000,000. This has been discussed for some time and passed the Senate without serious contest in early June. A few days later the other bill relating to the Farm Loan Act, increasing the interest rate to 5½ per cent, was adopted without serious opposition, although it was to be noted at this time that there was a growing feeling in the Senate that a new determination had broken out among the Senators from the South and West. With the coming of summer, there were frequent suggestions of adjournment.

On June 17th, final action was taken on the Packer Bill which was passed by a vote of 45 to 21 after it had been bitterly fought by those who objected to government supervision of the industry. During June there was almost constant debate on some agricultural measure and steady progress toward final action on numerous bills.

In July, however, came the first great struggle in which the Bloc was tested with respect to the sincerity of its interest in the agricultural program. The fact that this group had shown a determination to strive for action stimulated the opponents to agricultural matters to urge adjournment and when the suggestion was proposed it was necessary for the Bloc to stand solid against the idea of postponement and insist on enacting such relief measures as had been advanced to the point of passage. In discussing this question, Senator Kenyon said:

"I do not know what we can do in the way of an agricultural program. Some of us have been meeting and discussing the problem of what should be done practically, not any wild scheme, not any political nostrums that amount to nothing, but some real work to help the agricultural situation."

It will be noted in the above that Senator Kenyon referred to the Bloc meetings which some of the newspapers have endeavored to point out were secret conferences from the outset.

156   THE AGRICULTURAL BLOC

Among the arguments advanced for adjournment were the hot weather; the necessity of taking up tariff and revenue bills; and numerous other reasons principally designed to detract attention from the main objection which was to pass over the matter until some later date. Postponement is a favorite method of those who are opposing legislation and can usually be put down as a perfectly good reason for judging the speaker to be an opponent of a measure even though he will not admit it.

When the matter of adjournment came to a vote, the result was 27 to 24 against it, with 45 Senators absent or not voting. This was only a beginning, however, of a number of efforts to bring about an adjournment which made it necessary for the representatives of the Agricultural Bloc to stand solidly together at all times.

While June had been a month in which much legislation had been started on the way to passage, the record for August in the midst of one of our hottest summers was one of the best ever made so far as agricultural measures were concerned. During that month both the Packer Control Act and the Future Trading Act became law. The contest on the Future Trading

Act was much less than had been experienced with the Packer Bill, since by this time, August 9, the strength of the Agricultural Bloc had been recognized and the bill was finally adopted without a vote of record. The same was true of the amendment to the War Finance Corporation Act, which passed August 4 without a record vote.

When it was once apparent that the Bloc could mobilize the strength to secure final passage for a bill, the opponents of agriculture shifted their tactics to the extent that they began to present numerous amendments to various bills designed to modify the general purpose of the bill. The Bloc was just as effective in meeting these raids as it was in holding the Senate to its task and advancing the measures as rapidly as possible.

During the Fall there was a considerable period when the chief attention of Congress was directed to the tax bill and a discussion of tariff matters and relatively little attention was given to the measures on the program of the Farm Bloc. During this period occurred the contest over the Excess Profits Tax in which the Bloc was supposed to have functioned although, as a matter of fact, the group did not specifically regard the Revenue Bill as a part of its program.

With the opening of the new session in December, the agricultural program was again taken up with renewed vigor and the Capper-Volstead Bill on coöperative marketing was made the principal issue for immediate attention. This bill was particularly the target for objections by those who thought that it was entirely a special-privilege measure and gave undue privileges to a certain class—the farmers. A determined effort was made to so amend the bill as to render it valueless and it required the constant support of the Bloc to secure its passage. But this was finally accomplished in February by a close vote of 47 to 44. The bill providing for representation for agriculture on the Federal Reserve Board met a similar opposition, particularly from those representing financial interests, but was finally approved on January 17th by a vote of 42 to 38.

From then on through the session there were no important agricultural measures advanced. Most of the time was devoted to discussions of tariff.

Early in May at an open meeting the subject of Rural Credit legislation was discussed by the several Senators who had introduced bills, by representatives of the leading farm organizations and others. At this meeting a special committee was appointed to consider all the proposals and recommended a measure which the group could unitedly support.

This committee called in experts and made a very effective effort to reach a conclusion.

When the House adjourned for the summer recess on June 30, the Senate was considering the agricultural schedule of the tariff, which was adopted about as recommended by the agricultural tariff group to the Finance Committee.

A list of measures relating to agriculture, enacted since the Agricultural Bloc was organized is as follows:

Amendment to the Farm Loan Act, increasing the Capital $25,000,000.

Amendment increasing Farm Loan Bond Interest Rate to 5½ per cent.

The Future Trading Act.

The Packer Control Act.

War Finance Coöperation Amendment.

The Coöperative Marketing Bill.

Amendment to Include Representative for Agriculture on Federal Reserve Board.

The record of the Bloc movement in the House

160 THE AGRICULTURAL BLOC

of Representatives is not so easily shown, since the organization there has never been so clearly defined as in the Senate. About the time the Senate group was formed there was a meeting of a number of representatives from the House under the leadership of Representative L. J. Dickinson of Iowa. Committees were appointed on Transportation, Taxation and Revenue; Tariff, Credit and Miscellaneous measures. This group has not included all of those interested in agricultural measures, however, and can hardly be compared exactly with the Bloc in the Senate.

The results of organized interests in agriculture is shown by the fact that the House passed all of the measures enacted by the Senate by larger proportionate majorities than were secured in the Senate.

On several bills there was a hard fight by the farmers' representatives against those who endeavored to block or postpone legislative action. But after the first parliamentary skirmishing the final votes were large, as for example, 269 to 69 on the Future Trading Bill; 251 to 71 on the Packer Bill and 315 to 21 on the Amendment to the War Finance Corporation Act granting greater credits to farmers.

The same conditions that led Senators into group action also led the Representatives to act together. In the final analysis there is little difference between the reactions of a Representative and a Senator; if anything, the former is likely to think in terms of his home district more than the Senator who thinks in terms of the State and Nation.

The record of the Bloc includes all the list of measures, mentioned above, that have to do with agriculture in particular and the national welfare in general. We cannot say what might have been the situation had none of these bills been passed. We would still be discussing packer control, the anti-grain gambling bills and coöperative marketing. Instead we now have an efficient administration of these three acts in operation under the Secretary of Agriculture with good results already in sight.

The Bloc has undoubtedly speeded the progress of the movement toward better marketing and more justice in the distribution of the consumer's price between the producer and distributor. We still have far to go but we are on the way.

CHAPTER XV

WHAT THE FUTURE DEMANDS

THE future development of American agriculture has received too little attention. We have been so busy in the present-day work of subduing a naturally rich and productive continent that we have given little thought to the time when expansion will no longer be possible and a growing population will force us to seek new means of subsistence.

Other nations have met this issue through colonization, by sending their people into the far reaches of the world where unoccupied lands awaited settlement. National wealth has been developed horizontally by trade after the home country has reached a certain stage of development. But the era of expansion through colonies is approaching its end and the policies and disposition of the American nation have never been to look into the future from this point of view. Our whole idea of expansion has been to settle our own broad land and there has been plenty of room for all of us.

The next step that immediately presents itself is in expansion upwards, that is, by improved production on the acres now under cultivation by the application of better methods of farming, better seed, better live stock, better farm machinery, all directed by a thoroughly educated body of farmers. In this direction, there undoubtedly lie great possibilities and for the last fifty years the development of agricultural science has been making strides. This progress has been in the form of increasing production at a rate in keeping with the growth of population and the needs of industry. We have kept up our spirits by reciting constantly the story of the increase in production. We have boasted of the fact that since the Civil War we have increased our cotton production five times, our wheat production four times, our corn production three times, while our population only increased two and three-fourths times.

We have been gratified to know that we have produced one-fourth of all the wheat in the world, over half of all the cotton, and three-fourths of all the corn, with only one-sixth of the world's population.

But this is now a matter of history and while

the production of the American farmer exceeds that of any other farmer, per man, the peak of agricultural production per capita of population was passed in 1898, nearly a quarter of a century ago. To the uninformed, the answer to this check in the growth of production is further expansion, but we are face to face with the fact that our best lands are already under cultivation and expansion can occur only at the high cost of reclamation or the improvement of less desirable lands. The improvement in production through added acreage cannot alone keep pace with our needs.

We need to observe how highly concentrated is the production our principal crops. A recent survey of the five leading states producing certain crops shows the following interesting figures:

The leading five states for each crop produced in 1921:

66.8% of the Cotton Crop
78.4% of the Spring Wheat Crop
53.5% of the Winter Wheat Crop
42.4% of the Corn Crop
72.5% of the Market Apple
62.7% of the Barley Crop
42.4% of the Potato Crop
97.7% of the Flaxseed Crop

We also face new hazards not heretofore met. New countries have the minimum loss from the ravages of insects and plant diseases, but we

WHAT THE FUTURE DEMANDS 165

have reached the stage where the struggle between the farmer and these enemies is growing more vigorous day by day. New and heretofore unknown pests are developing every year that threaten to greatly limit if not at times entirely wipe out a large section of our food-producing area for particular crops. In many instances, pests have appeared which menace the production of staple crops heretofore regarded as entirely safe. An example is the corn borer, threatening to invade the area occupied by the richest food-producing crops we have. Another example is the fruit-fly, a constant menace to our fruit-growing resources. Other diseases in the soil and in the air add to the complications of the struggle and offensive battle which the farmer must bear. Science can survive the method of control in many instances, but only at great cost, and this cost must be borne by the consumers in the end.

It is conservatively estimated that by the close of this century the population of the nation will exceed 200,000,000 souls and the United States, which is to-day a food-importing nation, measured in dollars, will either have to depend more largely on imported foods or reorganize entirely our national life. We to-day import more in value of sugar, tea, coffee, spices and tropical fruits than we export of wheat and meat. While we will always continue to exchange our products for those to which our soil and climate is not adapted, the cost of that which we import must steadily rise.

These dangers are not of to-morrow or of the day after, but the time has come when we must develop a policy and plan for national growth which will preserve the balance between agriculture and industry. This policy must include every factor bearing on our national life and our relation with other nations. History has shown that only those nations who have given large attention to increasing the food supply in keeping with the increase in population, either through scientific production, such as Germany practised, or through trade as England has done, can expect to maintain a position of leadership.

The first plank in this platform must be the recognition by the entire nation that the soil is the foundation of all real wealth and only through fostering continued production from the soil can national growth be assured.

As a nation we have thus far spent comparatively little in recent years on the development and safe-guarding of agriculture. The entire cost of the Federal Department of Agriculture and all the educational institutions for agriculture, since they were first founded, is less than was spent during a few weeks of warfare by this nation alone in the recent war. And that war was the result, indirectly, of the pressure within a crowded continent for the opportunity to expand through trade with foreign countries.

These larger policies involve in their execution the handling of a multitude of smaller questions, one at a time, as we come to them in the course of our development. One by one, we must take up these minor questions and solve them in keeping with a sound permanent policy.

These questions may be ranged in the following brief form, for convenience in relating them; the order merely indicates their rank as they appear at the moment:

Financing agriculture comes first, since without a capital investment out of the national savings, sufficient to expand production, no great progress can be made. Our financing must involve not only the development and reclamation of all available lands, but particularly the equipping of farmers so that each may develop the maximum production on the area that he cultivates. We must have a system of farm credits which will place a dollar at every point in our production system where it will return more than a dollar in product. We must enable the young farmer to get the land and the equipment necessary to becoming a successful producer, and this involves the entire problem of land-mortgage-credit, financing a tenant, financing the production and movement of crops and live stock and its final distribution in the market.

Education takes second place in the rank alongside of financing. By education I mean not merely the training of farmers but the broader education involved in the encouragement of research and experiment which will develop new methods, improve existing crops and live stock, develop our knowledge of production and distribution, the handling of the products, standards of quality, the control of pests and diseases, and all of the many details that follow. Every dollar must be invested in scientific work that will bring a return of more than a dollar.

Transportation is the next fundamental, since transportation is vital to agriculture owing to the central fact that the consumer cannot be located near the point of production. This involves the development of our railways, high-

WHAT THE FUTURE DEMANDS 169

ways and waterways in keeping with the growth of the flow of products from centers of production to centers of consumption. We must understand that the preferred rate, if any, must be allowed upon the raw material which is vital to industry and consumers.

Improvements in transportation through the expansion of our railways, particularly, have been the means of bringing immense new areas under cultivation and in developing entirely new industries in agriculture. But the debacle of the last few years in our transportation system lays before us the enormous task of revising the haphazard systems into one smoothly working whole which will give the required service. Involved in this transportation question are not merely the problems of freight rates, car distribution and such details, but the bigger problem of the national investment in such services and the regulation of income to be derived therefrom both in the form of returns on money invested and the intangible returns in the form of service rendered to the people.

Distribution improvement, or *better marketing,* follows in the logical order though at the moment it stands at the top of the list requiring immediate attention to secure the proper distribution of the profit from production. The great inequalities in the division of the cost of the product to the consumer between the retail distributor, the handler, the transporter and the producer, grow out of our complicated system of distribution involving many expenses for service which cannot long endure. Improvement of our marketing system cannot be made at one stroke nor through the adoption of any single gigantic plan of replacement of the present system, but must come through gradual improvements, one step at a time, by eliminating excessive costs, regulating those who make unreasonable profits and charges and by avoiding waste. The wastes in our present distribution system will, if saved, more than pay in a single year for the cost of all the improved marketing devices yet adopted.

A *national policy of agricultural advancement* has been taking form rapidly during the last few years. Such a policy cannot be evolved in a day but must come gradually through the steady growth of knowledge and appreciation by large numbers of people. The movement which has taken the form of the subject of this book is in a way an expression of the growth of feeling among agricultural leaders. It has not been

WHAT THE FUTURE DEMANDS

merely a spontaneous outburst of enthusiasm on the part of a few, but the culmination of a steadily growing conviction on the part of that third of our people who live upon the land—the American Farmers.

THE END